DISCOVERY AND ISOLATION OF MICROBIAL PRODUCTS

ELLIS HORWOOD SERIES IN BIOLOGICAL CHEMISTRY AND BIOTECHNOLOGY

Series Editor: Dr. ALAN WISEMAN, University of Surrey

Ambrose, E. J.	**The Nature and Origin of the Biological World**
Berkely, R. C. W., et al.	**Microbial Adhesion to Surfaces**
Blackburn, F. and Knapp, J. S.	**Agricultural Microbiology***
Butt, W.R.	**Topics in Hormone Chemistry Vol. 1**
Dean, A. C. R., et al.	**Continuous Culture Vols. 6 & 8**
Dunhill, P., Wiseman, A. and Blakeborough, N.	**Enzymic and Non-enzymic Catalysis**
Horobin, R. W.	**Chemical and Physical Basis of Biological Staining***
Kennedy, J. F., et al.	**Cellulose and its Derivatives**
Kennedy, J. F. and White, C. A.	**Bioactive Carbohydrates**
Kricka, L. J. and Clark, P. M. S.	**Biochemistry of Alcohol and Alcoholism**
Palmer, T.	**Understanding Enzymes 2nd Edition**
Reid, E.	**Methodological Surveys in Biochemistry Vols. 6-11**
Rennie, M. J.	**Biochemistry of Muscle and Exercise***
Roe, F. J. C.	**Microbiological Standardisation of Laboratory Animals**
Sammes, P. G.	**Topics in Antibiotic Chemistry Vols. 1-6**
Sikyta, B. and Moss, M. O.	**Methods in Industrial Microbiology**
Verrall, M. S.	**Discovery and Isolation of Microbial Products**
Wiseman, A.	**Handbook of Enzyme Biotechnology 2nd Edition**
Wiseman, A.	**Topics in Enzyme and Fermentation Biotechnology Vols. 1-10**

*In preparation

DISCOVERY AND ISOLATION OF MICROBIAL PRODUCTS

M. S. VERRALL
Extraction Section Manager
Beecham Pharmaceuticals, Betchworth, Surrey, UK

Published for the
SOCIETY OF CHEMICAL INDUSTRY
London, by

ELLIS HORWOOD LIMITED
Publishers · Chichester

First published in 1985 by

ELLIS HORWOOD LIMITED

Market Cross House, Cooper Street, Chichester, West Sussex, PO19 1EB, England

The publisher's colophon is reproduced from James Gillison's drawing of the ancient Market Cross, Chichester.

Distributors:

Australia, New Zealand, South-east Asia:
Jacaranda-Wiley Ltd., Jacaranda Press,
JOHN WILEY & SONS INC.,
G.P.O. Box 859, Brisbane, Queensland 4001, Australia

Canada:
JOHN WILEY & SONS CANADA LIMITED
22 Worcester Road, Rexdale, Ontario, Canada.

Europe, Africa:
JOHN WILEY & SONS LIMITED
Baffins Lane, Chichester, West Sussex, England.

North and South America and the rest of the world:
Halsted Press: a division of
JOHN WILEY & SONS
605 Third Avenue, New York, N.Y. 10158 U.S.A.

British Library Cataloguing in Publication Data
Discovery and isolation of microbial products.
1. Micro-organisms – Identification
I. Verrall, M.S. II. Society of Chemical Industry
576 QR65

Library of Congress Card No. 85–16350

ISBN 0–85312–748–4 (Ellis Horwood Limited)
ISBN 0–470–20249–1 (Halsted Press)

Typeset by Ellis Horwood Limited
Printed in Great Britain by The Camelot Press, Southampton

Table of Contents

Preface

In recent years biotechnology has received much publicity, largely due to advances in recombinant DNA technology, and the layman could be excused for believing that biotechnology is a 'new', late-twentieth-century invention. Of course it has its roots in antiquity with the production of fermented foods and drinks, but throughout this century we have used the combined disciplines of microbiology, chemistry and chemical engineering in a formal manner to develop microbial products for other purposes.

In all cases the products, whether primary or secondary metabolites, are produced by the microorganism in an aqueous environment, usually in dilute solution and in the presence of a host of poorly defined contaminants. Therefore it is not just a matter of producing the required compound, we must also isolate it in a form suitable for its intended application. In the case of pharmaceutical products, purity is of paramount importance, whereas industrial enzymes, for example, can be used in a fairly crude form. The spectrum of microbial products covers such a vast range of chemical structures that it would appear almost impossible to generalise on isolation methods, and indeed a comprehensive treatise would consist of several substantial volumes.

The contents of this book are based on papers presented at the meeting 'Discovery and Isolation of Microbial Products' held in London on 23/24 October 1984. It was organised jointly by the Solvent Extraction and Ion Exchange Group and the Biotechnology Group of the Society of Chemical Industry. A two-day meeting can cover only a selection of topics, but the aim was to give broad coverage of microbial product isolation in terms of techniques rather than specific products, although a conscious decision was taken to exclude 'bulk' microbial products such as ethanol.

All chapters are of an essentially practical nature and should be of value to newcomers in the field. There is no diminution in the number of reports of novel, biologically active, secondary metabolites. It follows therefore that the process of distinguishing novel from known compounds is becoming more demanding and, to avoid unnecessary duplication of effort, a comprehensive data base of known compounds is required. The first chapter describes the origin and use of the most comprehensive such data base generally available, that of János Bérdy. Recent advances in analytical techniques have provided the means for rapidly elucidating gross structures or at least structural features of complex molecules, but it is normally necessary to obtain a pure specimen of at least a few milligrams. Before identification, possibly the only known property of the compound is its biological activity and interpretation may be confused by mixtures or by synergistic or antagonistic effects. Some initial approaches for isolating the compound are outlined by Steven Box and further scale up of specific techniques is described in the chapters by Walter Voser, Marico Pirotta and Anders Törnqvist. Microbially produced macromolecules are currently described more in terms of biological properties than of precise chemical structure. However, some chemical properties, for example, stability, are of great importance in designing a purification procedure. The chapter by Winkler, Wiseman and Sadler describes a procedure for isolating the unstable cytochrome P450 enzyme. Another factor of increasing importance, particularly for rDNA protein recovery, is cost. The methods of Per Hedman, of Howard Chase and of Lowe and Stead provide complementary low-cost procedures for isolating microbial proteins. The extension of rDNA technology to downstream processing is described by Alan Thomson and by Fullbrook and Parker. When scaling up any process, microbial, biochemical or chemical, safety becomes an increasingly important aspect. Roger Nourish and Nigel Brace provide the viewpoint of the Health and Safety Executive. Although only relevant in statutory terms in the United Kingdom, their philosophy surely has universal application.

Returning to the Meeting, I would like to thank Jane Bovier, Margaret Johnson and the Conference Secretariat of the Society of Chemical Industry for their efficient administration, Dr M. J. Slater, Mr R. M. Carlyle and Dr M. K. Turner for chairing various sessions, and Mr C. J. Davis for editorial assistance.

M. S. VERRALL

1

Screening, classification and identification of microbial products

János Bérdy
Institute for Drug Research, Budapest, Hungary

1.1 INTRODUCTION

The topic selected for this chapter is so broad that, considering the limited space available, I have decided to present only schematically the classification problems of microbial metabolites, especially antibiotics, and in the second part some specific aspects of the screening and identification of bioactive metabolites.

Antibiotics, the most important type of microbial metabolite, represent one of the largest and most diverse groups of natural products. Most antibiotics are produced by microorganisms such as actinomycetes, bacteria and fungi, but antibiotic compounds may be obtained from many other living organisms such as higher plants, lichens, algae and animal organisms (Table 1.1). The antibiotics represent in practical terms a very important group of compounds used in the therapy of infectious and neoplastic diseases. It is not wishful thinking to predict that microbial products will also gain importance in the therapy of various non-microbial diseases. In addition antibiotics are widely used in veterinary medicine, agriculture, food preservation, biochemical research and several other fields.

Table 1.1 – The number of known antibiotic compounds (1984).

Type of producing organism	Approximate number	Percentage
Bacteria	950	9
Actinomycetes	4600	43
Fungi	1600	15

Table 1.1 (cont.)

All microorganisms	7150	67
Lichens	100	1
Algae	250	2
Higher plants	2500	23
Animal organisms	700	7
All 'higher' organisms	3550	33
Total	10700	100

I remember that, almost a quarter of a century ago when I started to work in this field, the number of antibiotics known at that time was only a few hundred. This number has increased today to more than ten thousand including about seven thousand microbial metabolites (Bérdy 1980). In the last decade, in addition to antibiotics about two thousand further bioactive secondary metabolites, exhibiting the most diverse bioactivities (enzyme inhibition, physiological effects, etc.) have also been isolated from microorganisms (Umezawa 1982). Estimation of the number of other microbial metabolites without any recognised bioactivity is very difficult, but their number is surely over fifteen thousand.

1.2 CLASSIFICATION OF ANTIBIOTICS

1.2.1 Survey of classification methods

The classification of microbial metabolites having bioactivity (for example antibiotic activity) is possible from several different points of view. Physicians, clinicians and pharmacologists want a classification according to the biological activity of the compounds; biochemists, molecular biologists and theoretical clinicians want to classify according to mechanism of action. Microbiologists prefer a classification according to the producing organism, that is, on the basis of its origin, or a system based on the biosynthetic pathway. Chemists first of all want to classify the isolated compounds according to their physicochemical properties or more rationally on the basis of their chemical structures.

Any kind of classification has some advantages and disadvantages (Bérdy 1974; Kurylowicz *et al.* 1975). Classification according to bioactivity has practical value for grouping medically useful antibiotics and those applied in other fields. There are: medical and agricultural antibiotics; antibacterial and antifungal compounds; 'narrow' and 'broad' spectrum antibiotics; compounds active only against Gram positive bacteria and those which are active against Gram negatives and Mycobacteria, also antitumour, antiviral, antiprotozoal, antimycoplasma and antiparasitic compounds. These activities sometimes may occur alone or more frequently in various combinations. Again, there are compounds which are active against only one or just a few particular organisms (e.g. bacteriocins, saramycetin, etc.), while others have a very broad activity range (e.g. streptothricins). The agricultural antibiotics are used as feed additives or plant protectants against bacterial, fungal and viral infections. Others are applied as preservatives or biochemical reagents. Classification according to bioactivity would be rather vague owing to the poor and uneven characterisation of many compounds, and it would be devoid of precise scientific value.

Rational classification based on the mechanism of action is difficult in the light of the large number of known antibiotics. The exact mechanism of action is known for only a few hundred antibiotics. An individual antibiotic sometimes has different types of molecular mechanism and often it is difficult to identify the primary and secondary processes. Completely different chemical structures frequently exhibit similar or identical mechanisms. The *in vitro* results do not necessarily reflect the *in vivo* mechanism. This classification would have to be revised frequently according to our increasing knowledge of molecular biology.

Classification according to the origin of these biologically derived compounds permits the establishment of several large groups, but rational subdivision is almost impossible within the framework of this system. There are organisms such as certain *Streptomyces*, *Bacillus*, *Pseudomonas*, or *Penicillium* species which produce a large number – sometimes 50 to 60 – antibiotics with diverse bioactivity and chemical structures. On the other hand, the same antibiotics as a rule are isolated from many different species of the same genus. Oxytetracycline for example is produced by more than twenty different *Streptomyces* species. There are many antibiotics which are produced by totally different micro-organisms. The best known example is the production of several β-lactams by *Streptomyces* species, fungi and bacteria. It seems, however, that generally the production of a particular type is more or less characteristic of a given class or genus. *Bacillus* species produce almost exclusively peptide derivatives; but macrolides and polyenes are almost exclusively *Streptomyces* and *Micromonospora* products.

Classification according to biosynthesis already has a significant biochemical background. Nowadays several hundred antibiotics have known biosynthetic pathways, and the biosynthetic route is presumed for many other compounds. Similar chemical structures are generally formed by similar biosynthetic pathways, but very frequently the same biosynthetic route results in different chemical structures with very different bioactivities. The frequently occurring mixed biosynthetic pathways also restrict rational and unambiguous classification on the basis of this principle.

Classification of antibiotics according to physical and chemical properties for example on the basis of elementary analysis, optical rotation, UV or IR spectra, functional groups, molecular weight, etc., has considerable practical value. Classification according to the solubility and acid–base characteristics, into hydrophilic or lipophilic, and into acidic, basic, neutral or amphoteric compounds, those that can be extracted from the filtrate by various solvents at different pH values, or can be absorbed on ion exchange resins or on other sorbents, has particular importance in screening and isolation processes. Consideration of the above data together with chromatographic data using card index systems or computers, strongly facilitates an effective antibiotic development programme. This type of classification, however, lacks precise scientific value.

In summary, in the case of many compounds today our knowledge is insufficient as regards their mechanism of action or biosynthesis for a detailed classification. Subdivision in a system based on the origin or bioactivity in most cases is only approximate. Our insufficient or frequently changing knowledge, or the inherent feature of some classification principles, limits the possibility of classification according to the above mentioned principles. However, in the case of the chemical structure of compounds, we already have sufficient characteristic information to establish a classification system of both practical and theoretical value (Bérdy 1972, 1974).

1.2.2 Chemical classification

The chemical structure of compounds is the property that unambiguously determines their physical, chemical, biological, pharmacological, and eventually their clinical properties. Owing to the great development of physical methods for structure elucidation, today the majority of new antibiotics have known chemical structures and more and more of the structures of previously isolated and poorly characterised 'old compounds' are also being elucidated. Additionally, there are numerous compounds about which we have sufficient information on their molecular skeleton, to allow at least a rough primary classification (Bérdy 1970; Round Table Discussion 1970).

Antibiotic structures express features of the entire range of organic chemistry. It is one reason why it is, in contrast to the alkaloids, steroids, terpenes or enzymes, extremely difficult to achieve a homogeneous chemical classification of antibiotics.

The classification system which I proposed more than ten years ago (Bérdy 1972) follows formal chemical classification, but not in the strictest sense. Because antibiotics are biosynthesised from simple organic molecules such as sugars, amino acids, fatty acids, purine–pyrimidine bases, acetate–propionate units, it is obvious that these moieties or skeletons derived from them should represent the first basic principle of classification.

There exist additionally some specific moieties that occur almost exclusively in antibiotics and play an outstanding role in their bioactivity. These constituents, for example macrocyclic lactone or lactam rings, cyclic peptides, β-lactam units, some specific quinone-like structures, deoxystreptamine, etc., have to be considered with due emphasis during classification. These finally selected simple molecules and specific moieties, the 'principle constituents' playing an important role in the structure and activity of antibiotics, form the basis of primary classification. These principal constituents, representing the nine antibiotic families, and the 35 sub-families corresponding to these specific moieties are listed in Table 1.2.

Table 1.2 – Key to the main antibiotic families.

Primary code numbers	
1	CARBOHYDRATE ANTIBIOTICS
1.1	Pure saccharides
1.2	Aminoglycoside antibiotics
1.3	Other (N- and C-) glycosides
1.4	Various sugar derivatives
2	MACROCYCLIC LACTONE (LACTAM) ANTIBIOTICS
2.1	Macrolide antibiotics
2.2	Polyene antibiotics
2.3	Other macrocyclic lactone antibiotics
2.4	Macrolactam antibiotics
3	QUINONE AND SIMILAR ANTIBIOTICS
3.1	Linearly condensed polycyclic compounds
3.2	Naphthoquinone derivatives
3.3	Benzoquinone derivatives
3.4	Various quinonelike compounds
4	AMINO ACID, PEPTIDE ANTIBIOTICS
4.1	Amino acid derivatives
4.2	Homopeptides
4.3	Heteromer peptides
4.4	Peptolides
4.5	High molecular weight peptides
5	NITROGEN-CONTAINING HETEROCYCLIC ANTIBIOTICS
5.1	Noncondensed (single) heterocycles
5.2	Condensed (fused) heterocycles
5.3	Alkaloids with antibiotic (antitumour) activity
6	OXYGEN-CONTAINING HETEROCYCLIC ANTIBIOTICS
6.1	Furan derivatives
6.2	Pyran derivatives
6.3	Benzo[γ]pyran derivatives
6.4	Small lactones
6.5	Polyether antibiotics
7	ALICYCLIC ANTIBIOTICS
7.1	Cycloalkane derivatives
7.2	Small terpenes
7.3	Oligoterpene antibiotics
8	AROMATIC ANTIBIOTICS
8.1	Benzene compounds
8.2	Condensed aromatic compounds
8.3	Nonbenzoid aromatic compounds
8.4	Various derivatives of aromatic compounds

Table 1.2 – *Continued.*

Primary code numbers	
9	ALIPHATIC ANTIBIOTICS
9.1	Alkane derivatives
9.2	Aliphatic carboxylic acid derivatives
9.3	Aliphatic compounds with S or P content
0	MISCELLANEOUS ANTIBIOTICS (with unknown skeleton) (May be distributed to physical, chemical and microbiological properties)

In the case of more complicated compounds, sometimes there is more than one principal constituent in the molecule, so it is obvious that the same compound may belong to different classes. To avoid this unnecessary duplication, the most characteristic principal constituent was selected and the primary classification was made accordingly. In this selection mainly biogenetic points of view and several physical, chemical, and biological similarities were taken into account. The existing second and perhaps third most important principal constituents represent the basis of secondary classification. The secondary and tertiary classification was done individually on the basis of existing other constituents or specific moieties, the size of molecule (e.g. amino acid, oligo- or

Table 1.3 – Classification of antibiotic compounds.

AN	Family, sub-family, group	Important representatives
1	Carbohydrate antibiotics	
11	Pure saccharides	
111	Mono and oligosaccharides	Streptozotocin, nojirimycin
112	Polysaccharides	Glucans, soedomycin
12	Aminoglycoside antibiotics	
121	Streptamine derivatives	Streptomycins, bluensomycin
122	2-Deoxystreptamine derivatives	Neomycin, gentamicin, etc.
123	Inositol-inoseamine derivatives	Kasugamycin, validamycin
124	Other aminocyclitols	Fortimicin
125	Aminohexitols	Sorbistin
13	Other glycosides	
131	Streptothricin group	Streptolin, racemomycin
132	Glycopeptides, C-glycosides	Vancomycin, chromomycin
14	Sugar derivatives	
141	Sugar esters, amides	Everninomicin, lincomycin
142	Sugar lipids	Moenomycin, labilomycin

poly-peptide, protein), the nature of the chromophore, molecular skeleton, specific linkages, physicochemical and biological similarities, etc. Subdivision of the carbohydrate, antibiotic family is illustrated in Table 1.3.

Since this system represents merely an artificial classification, the construction of more or less arbitrary categories of compounds seems to be necessary; but it is justified by the requirement of homogeneity. Limiting the carbohydrate family of antibiotics to the mainly sugar-containing structures, excluding most of the glucosides such as macrolides, anthracyclines, nucleosides and aromatic glycosides, which are classified according to their different aglycones, surely contributes to a logical classification. Formation of the macrocyclic lactone family and the separation of the quinones from the true – mainly phenolic – aromatic compounds, was necessary. The frequent natural occurrence of macrocyclic lactones and quinone compounds, their practical importance (macrolides, ansamycins; tetracyclines, anthracyclines) as well as their completely new biological and chemical properties, different from those of aliphatic and aromatic antibiotics, justifies their listing as a separate family of antibiotics. Arbitrary decisions became necessary in the grouping of streptothricins, glutarimides, alkaloids and a few sulphur-containing heterocyles, but I think these do not affect the inherent, logical arrangement and usefulness of this system.

To facilitate orientation among the compounds, and for easy retrieval and computerisation, a numerical classification was introduced (Bérdy *et al.* 1980–82). Every entry (every chemical entity) has an Antibiotic Classification or Antibiotic Code Number (AN). The first element of this five-digit number represents the antibiotic family, the second one the sub-family, and the third, fourth and eventually fifth number refer to the group, type and sub-type respectively (Table 1.4). Closely related compounds have identical classification numbers. Poorly characterised compounds receive only the first elements (completed with zeros). The AN itself, therefore, represents information about the chemical structure of the antibiotic and the depth of the knowledge about this compound. Competing the AN with a simple sequence number gives the Compound Number (CN) formed from the Code Number and Sequence Number, representing a numerical title for every compound.

This classification system divides the antibiotics into about 500 types and sub-types. Only 10–15% of all known antibiotics currently have insufficient data for characterisation and have to be classified into the miscellaneous group of compounds (AN: 00000) with unknown structural features. A feature of this kind of classification is the possibilty of creating an almost infinite number of new sub-families, groups, types, or sub-types for the new compounds which have yet to be recognised. In the last 10–15 years years no entirely new type of antibiotic has been discovered which could not be placed into the existing families of sub-families. A few new groups (e.g. combimicins, monobactams, etc.) and numerous new types and sub-types were added to the system (e.g. trioxacarcins, myxovirescins). It is essential of course to revise the system periodically with respect to newly recognised or modified chemical structures.

With the introduction of numerical classification, the system was ready for computerisation. The processing of my longstanding antibiotic data bank compiled on a card index system (Bérdy 1961) into computer-searchable format enabled the publication of this data bank in a series of Handbooks published by the CRC Press Inc., Boca Raton, Florida, USA. The first ten volumes of this *Handbook of Antibiotic Compounds* series covering 6700 antibiotics is already published (Bérdy *et al.* 1980–82). The next two volumes including 2100 new compounds and much new and revised data for old antibiotics were prepared in the last two to three years, and published in 1985. I am at present working with the next two volumes which will also cover about 2000 new antibiotics and new data.

Table 1.4 – A numerical classification.

1	Carbohydrate antibiotics	Family
12	Aminoglycosides	Sub-family
122	2-Deoxystreptamine glycosides	Group
1222	4,6-Disubstituted derivatives	Type
12221	Kanamycin type	Sub-type
12222	Gentamicin type	
⋮	⋮	

12222-3866 Gentamicin A_3

Code number Sequence number

1.3 THE SCREENING PROCESS FOR BIOACTIVE MICROBIAL METABOLITES

1.3.1 General considerations

Research into new drugs may be divided conveniently into two preclinical phases: discovery and development. The screening of microbial cultures for their antimicrobial, antitumour or other activities corresponds to the discovery phase of new drug research, and the evaluation of the new compounds obtained corresponds to the development phase. The discovery of microbial metabolites has several specific features, and it is more complicated than the search for new synthetic drugs (Hamill 1977).

An antibiotic development programme or any programme devoted to isolation of bioactive microbial metabolites depends on the combined results obtained from a whole technological evaluation team. It involves the contribution of scientitists of many different disciplines who make special and important contributions in this complicated project (Table 1.5). The table also illustrates the main stages of the project.

Table 1.5 – Contributors to the screening project.

Microbiologists:	Culture isolation and selection Cultivation (fermentation) Detection (screening) Yield improvement, preservation Biosynthetic modifications
Chemists:	Isolation and purification Identification Structure elucidation, synthesis Chemical modification
Pharmacologists: (Toxicologists)	Evaluation (*in vitro* spectrum of activity) Animal tests (*in vivo* evaluation) Toxicology Other bioactivities, side effects Pharmacokinetics (absorption, distribution, elimination, etc.)
Biochemists: (Molecular biologists, Geneticists, Enzymologists)	Biogenesis Mode of action Mutation, resistance, etc.
Clinicians: (Veterinarians, Agronomists, Nutritionists)	Clinical and field trials Pharmacokinetics in man, side effects Interactions, combinations, etc.
Bioengineers:	Scale up (biotechnology, yield improvement) Quality control
Pharmacists:	Formulation

The flow chart of the screening process from the isolation of the producing microbe to the recognition of a new active metabolite (the discovery phase of the research project), is illustrated in Fig. 1.1. This research phase normally takes at least one year for a culture to be isolated, have its activity reconfirmed, and be evaluated for novelty. Depending on the screening method applied and the type of producing species investigated, the screening of about 100 000 soil microbes may lead to only 5 to 50 new compounds, and there is no guarantee that from these new compounds, after detailed evaluation, we will find any new drug or other practically useful compound.

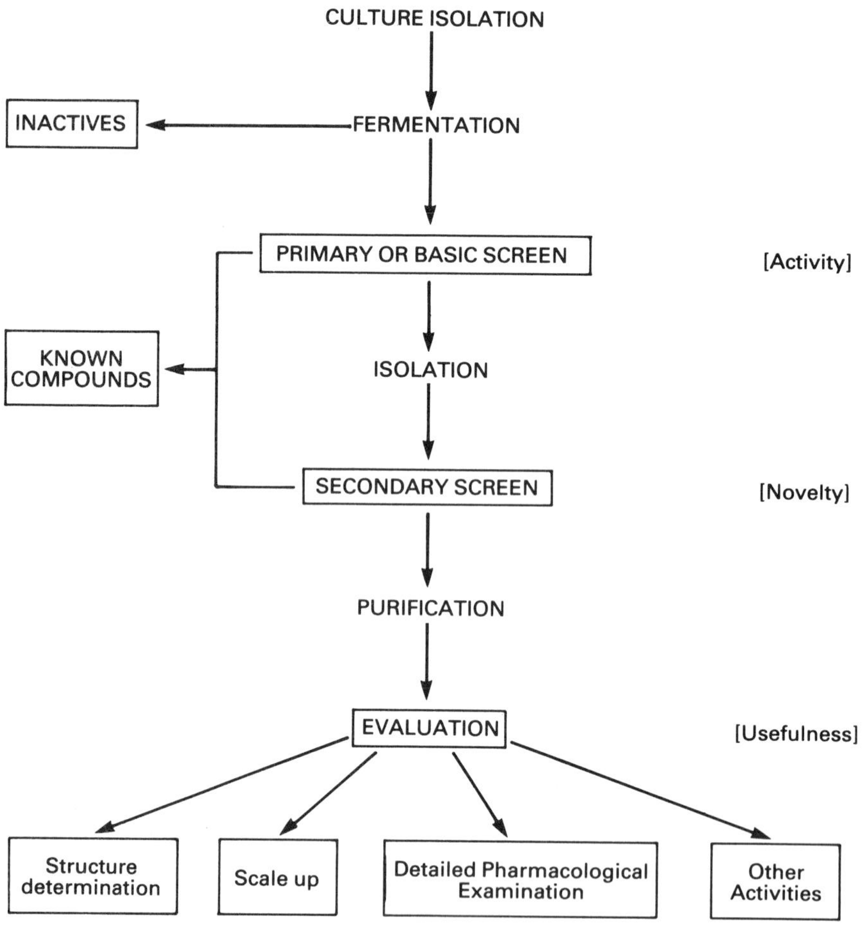

Fig. 1.1 – Screening process for acquisition of a new bioactive microbial metabolite (drug discovery phase).

After the recognition of a new microbial metabolite with promising bioactivity, its evaluation (the drug development phase) if we are lucky leads to one or a few promising agents which satisfy the requirements of the clinical trial. This phase of research is illustrated in Fig. 1.2.

For detailed evaluation of the newly discovered metabolite including animal tests, toxicology, pharmacokinetics, scale up, structure determination and modifications, investigation of other activities, etc., 2–3 years are required. The clinical or field trial of a selected compound normally takes an additional 3–6 years. Therefore, the development of a new microbial drug from culture isolation to the market place, requires 8–10 years, and at least 10–15 million dollars. In the following I do not want to touch the financial and economic problems derived from the above facts, rather I will discuss some specific features of the screening process.

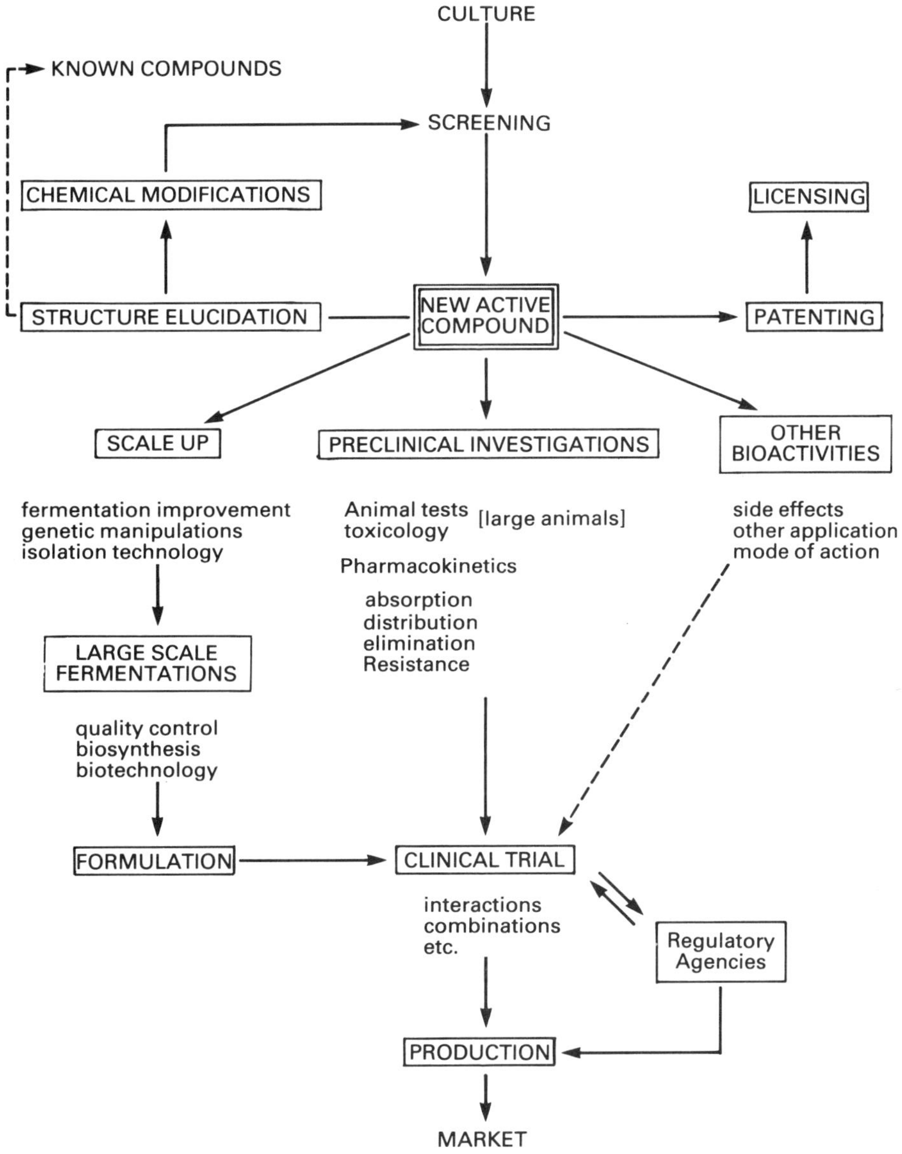

Fig. 1.2 – Evaluation of a new active compound (drug development phase).

1.3.2 Screening

The detailed route of a fermentation product from the isolation of producing microbe to the recognition of a new, promising compound is summarised in Fig. 1.3. This outline scheme is of course only one idealised possible way to obtain new microbial metabolites.

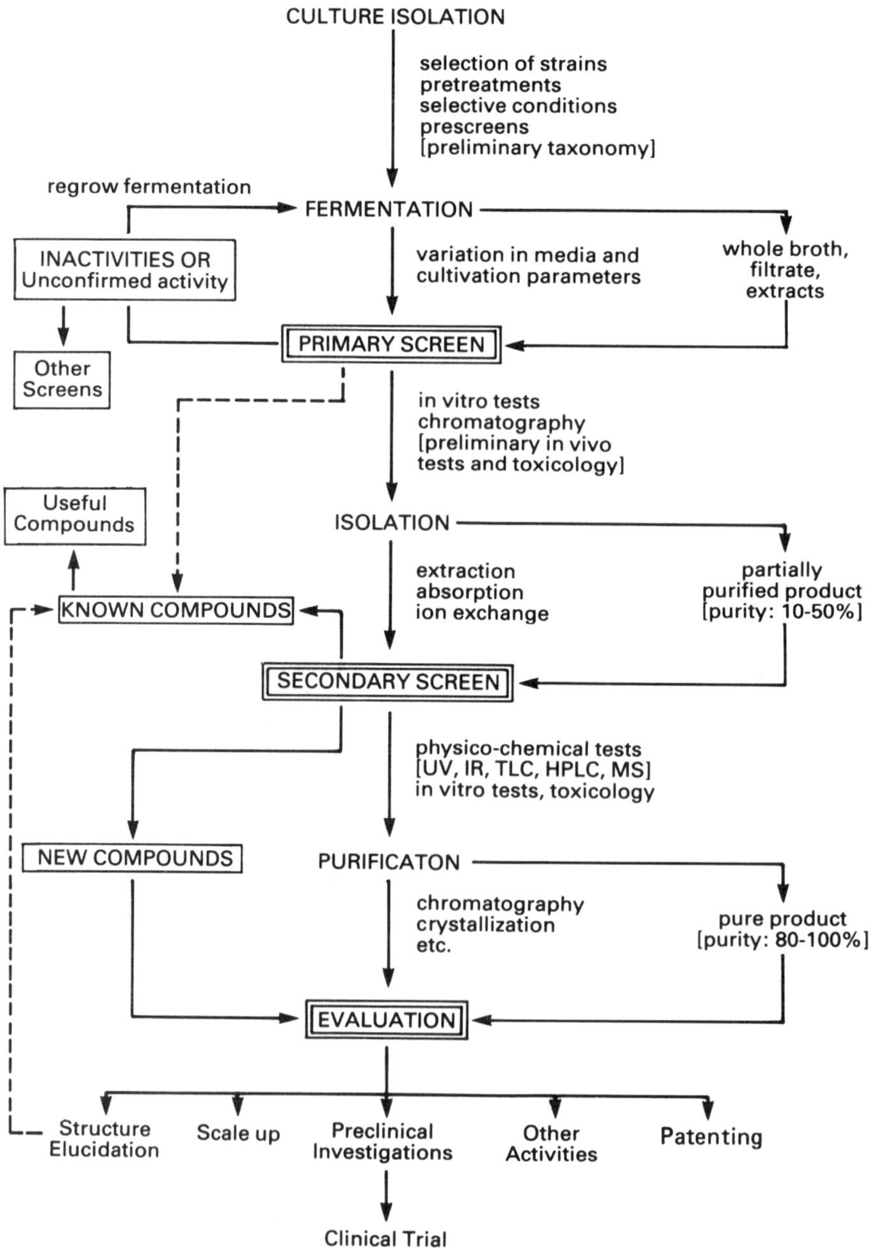

Fig. 1.3 – Detailed pathway of a new bioactive metabolite.

In practice there are no two identical screening projects. In some programmes, especially in the antitumour screening projects, various prescreens (e.g. *in vitro* test, cytotoxicity) precede the basic *in vivo* antitumour screen. The range of activities expected in the first stage, is based on the uniqueness of culture, novelty of the specific activity etc. Some laboratories do taxonomic studies first and screen only unique cultures for example rare actinomycetes (*Actinomadura*, *Actinoplanes* or *Micropolyspora* species), and other specific organisms such as Myxobacteriales, or *Mycoplasma* species.

Sometimes no isolation work is done before the detailed *in vivo* or toxicological evaluation. However, animal tests are best performed with purified products owing to the generally low concentration of the active compound and possible interfering effects from other compounds present in the broth or crude extract. In this case close cooperation and understanding between sample submitter and the pharmacologists may help the correct evaluation of the activities.

One of the greatest barriers to the discovery of an active metabolite with any kind of *in vitro* or especially *in vivo* screen, is sensitivity. The extreme dilution of the active material in the broth or crude extracts (sometimes 0.1–10 mg dm^{-3}) results in activity being observed only when the compound is extremely potent. Many active materials are undetectable, and those detected tend to be discovered repeatedly. Sometimes, therefore, fermentation improvement studies and genetic mutational programmes can be initiated to improve the yield of material before detection, purification and evaluation.

Preliminary identification of known compounds or commonly occurring undesired chemical classes (e.g. streptothricins, anthracyclines or polyenes) is possible in the early phase of research; but the certain elimination of all the known compounds in most cases is possible only after some isolation and purification work. Closely related analogues can be recognised only after detailed investigation of isolated pure substances. Other less interesting compounds may be worked up when time permits or sent to other screening institutions.

The most important criterion for judging any screen system is its capacity to detect novel compounds with potential utility, and its rejection of known active agents. In some specific situations, however, for patent or licence purposes, it may be useful to find new and perhaps more effective producers for known active agents, especially in the developing countries.

In general, despite automated techniques and instruments, I think one of the most important aspects of screening which will not be changed by any further development or automation, is the participation of an experienced investigator. The intuition, instinct and curiosity of microbiologists, chemists and pharmacologists are absolutely necessary for successful research.

When the novelty of an isolated compound is proved, additional *in vivo* testing, further biochemical and toxicological characterisation as well as structure determination of the unknown compound will provide enough information for a rational decision to be made as to whether the compound should be developed further. In the case of a favourable decision, the original fermentation parameters, isolation procedures, biological assay method etc., can be modified. Further strain selection, mutation programmes, optimisation of fermentation conditions, precursor feeding and other modifications can dramatically increase yields, sometimes 10- to 50-fold or more.

If the isolated compound represents a new or unusual type, analogues can be prepared by chemical or biosynthetic modifications. These transformations, sometimes result in stable or more active derivatives especially if the parent compound is of low activity or unstable as in the case of rifamycins and thienamycin.

To summarise, the ideal screen should be rapid, sensitive, inexpensive and suitable for high throughput, and would select only those compounds which have clinical activity. There is no single screening programme, however, capable of detecting all compounds desired. In reality every screen will miss some useful activities (false negatives) and some of the selected compounds will turn out to be inactive (false positives). Of course the majority of compounds, do not have the desired activity (true negatives).

1.3.3 Screening strategy

For planning an effective screening project both scientific and logistic decisions are required in balancing the numbers and types of organisms selected, the prescreens and basic screens applied, the fermentation conditions and number and type of test organisms or screening models used. One of the oldest questions of screening strategy is constantly in the foreground: is it better to screen a large number of cultures by a single screen and concentrate only on the major component of the particular broth? Or might it be better to study exhaustively a few cultures selected carefully and discover every type of activity and every metabolite produced by these few cultures? Or should both be done together? The limitations in manpower naturally preclude this third possibility. Mass screening methods using various autoanalytical techniques are still used in many pharmacuetical companies, but it has to be considered that today at least millions of organisms must be screened using conventional techniques to obtain a single utilisable new drug. Numerous very useful cultures, e.g. cephalosporin C or clavulanic acid producers, have not come from mass screening studies. Many cultures discarded over the years could produce numerous very interesting new types of compound. Reinvestigation of the exhaustively studied Actinomycetes employing new improved methods, therefore, may provide many interesting new specific activities.

On the other hand, modern screening methods, for example the screening on the basis of mode of action or on an enzyme inhibition concept, are not a satisfactory solution either. Activity in a cell-free macromolecular system does not necessarily mean activity in whole cells, and even less so in animals. It is a fact, however, that sophisticated exhaustive investigation of producing species and the use of new screening techniques have more frequently led to interesting new type of compounds of lower toxicity.

1.3.4 New screening methods

The two key features of the whole screening process are:

(i) the sensitivity and selectivity of the basic screen method applied

(ii) the early recognition of known and undesired compounds.

The basic screen used in the programme, i.e. the test organism or the *in vitro* or *in vivo* detecting system applied, is the determining factor in the efficacy of the whole screening process. Productivity of a conventional screening process based on the long-term exclusive utilisation in the primary screen of common test organisms – connected or unconnected with human diseases (*Bacillus subtilis*, *Staphylococcus aureus*, *Escherichia coli* and *Candida albicans*), has been greatly reduced in the last decade.

Rediscovery of compounds had many years ago already reached an alarming 90–95%, especially in the screening of soil *Streptomyces*. The introduction of automated techniques in strain isolation, cultivation and detection serves only to increase the number of strains investigated, but fails to affect the efficiency of the screening itself.

Introduction of new principles instead of random, mechanical *in vitro* screening is an absolute requirement for rejuvenating microbial metabolite research (Ninet *et al.* 1981; Zähner 1978). One might wonder why it took almost 25 years to discover the new β-lactam antibiotics, which are produced by a wide range of *Streptomyces* and other species. It is certain that these compounds were present on many occasions in the broths investigated but, owing either to their low *in vitro* activity, their instability or perhaps their low concentration, they were not detected in the conventional processes. Introduction of hypersensitive mutant test organisms and the development of β-lactamase inhibition tests were required for their discovery. These tests and several recently introduced new screening methods are listed in Table 1.6. Laboratories with a wide range of interests had already 10 to 20 years ago supplemented classical *in vitro* screens with the use of resistant and more specific test organisms such as those of recent clinical interest (e.g. anaerobes, facultative pathogens), hypersensitive mutants and several simple *in vivo* models. Of course it is not necessary in the primary screen to choose growth inhibition as an indicator of a useful metabolite. Application of a series of new methods based on entirely new principles such as antimetabolite, promoter–potentiator screens, inhibition of spore or sphaeroplast formation, induction of irregular growth of fungal mycelium, and the large variety of pharmacological and other biochemical tests are representative of these new methods.

Table 1.6 – Methods for screening of bioactive secondary metabolites.

Antimicrobial/antitumour tests:

Resistant strains, anaerobes, facultative pathogens
Specific test organisms (*Streptomyces*, *Clostridium* sp.)
Hypersensitive mutants
Antimetabolite tests
Cytotoxicity (KB, P-388, L-1210 cell cultures)

Table 1.6 (cont.)

Phage/prophage induction, antiphage activity	
Various (e.g. respiration deficient) mutants	
Lysogenic induction, 'BIP' test, mutagenesis	
'*in vivo*' models	
Other bioactivities:	
Inhibition of germination, spore or sphaeroplast formation	
Irregular growth of fungal mycelia	
Mitotic inhibitors	
Promoter–potentiator screen	
Inhibition of chemotaxis or phototaxis	
Immunity-enhancing compounds	
Cell-free systems:	
Inhibition of cell wall biosynthesis	mode of action screens
protein biosynthesis	mode of action screens
RNA or DNA biosynthesis	mode of action screens
fatty acid biosynthesis	mode of action screens
Membrane permeability tests	
Enzyme inhibition (β-lactamases, proteases, sialidases, etc.)	
Pharmacological actions:	
Chemical systems:	
Specific chemical reactions	
Chromatographic (TLC, PC, HPLC) methods	
Incorporation of radioactive isotopes	

The design and development of *in vitro* screens which are specific for the detection of a particular mechanism of drug action is the most challenging task, particularly considering the multitude of potentially interfering compounds which may be present in the broth or crude extract. Systems based on the mode of action – namely on protein-, RNA-, DNA-, or cell-wall biosynthesis inhibition – and other cell-free systems, for example membrane permeability models, are all useful new approaches in the discovery of new antibiotics with a new type of activity, lower toxicity and *in vivo* activity against clinically important pathogens without development or resistance.

A different approach, screening on the basis of chemical reactions, chromatographic properties, or incorporation of radioactive sulphur or halogens, helps us to find with minimum time and cost potentially useful new metabolites.

A very promising new approach is the inhibition of various enzymes of antimicrobial, agricultural or physiological interest. Inhibition of β-lactamases, proteases, peptidases, hydroxylases, sialidases or chitinases, some of them playing a role in cell-surface phenomena, have greatly expanded the scope of screening methods and the range of potentially useful new compounds.

Since there are over one hundred types of human cancer it is quite unlikely that a single screening model would be predictive of clinical activity in the antitumour area (Douros 1976). Owing to this diversity and specificity it is understandable that tumours of other mammals may not adequately respond to the compounds which would be effective in humans. Today it seems that useful antitumour compounds with a specific mode of action and lower toxicity could be more easily detected with various biochemical screens than with the usual *in vivo* animal models or even *in vitro* cell lines (Umezawa 1984). Antimetabolite screens, detection of potential activities involving the enzyme inhibition concept, the search for immunity-enhancing compounds and mitotic inhibitors, or blocker of nucleic acid synthesis, as well as agents effective on some cell-surface phenomena are all relatively new and promising approaches.

In the field of microbial metabolite research, a new horizon was opened 10 to 20 years ago when the systematic research of small molecular enzyme inhibitors and other bioactive metabolites started (Umezawa 1972). This research line has resulted already in almost two thousand new metabolites such as bestatin, coformycin, krestin, pepstatin, compactin (mevinolin), and various glycosidase inhibitors (Hamill 1982). It was proved that *Streptomyces* species and other microorganisms are all rich sources of enzyme inhibitors and other secondary metabolites with diverse biochemical and physiological activities (Table 1.7).

Table 1.7 – Secondary metabolites.

Chemical features:	Bioactivities:
Saccharides Lactones, lactams Pigments (quinones, anthrocyans) Peptides Nucleotides Alkaloids Flavonoids Terpenoids Aromatics etc.	Antimicrobial (antibacterial, antifungal, antiprotozoal) activity Antiviral activity Antitumour (cytotoxic) activity Insecticide, miticide, nematocide, acaricide, anthelminthic activities Toxic effects (phytotoxins) Growth-regulating and growth-promoting, hormone-like activity Pesticide, herbicide effects Immunity-enhancing action Ionophoric activity Enzyme effect Enzyme-inhibiting activity Physiological (cholesterol lowering, hypotensive etc.) activities etc.
Unusual structures: β-lactams, macrocycles, polyenes, polyines, multicondensed aromatics	
Unusual groups: Covalently bound halogens, nitro, nitril, isothiocyano, phosphino, phosphono groups, etc.	

The systematic search to obtain various bioactive compounds proved that a much larger variety of low molecular weight bioactive compounds could be isolated from microbes than was formerly believed. It is most likely that all secondary metabolites of microbes have some kind of bioactivity, which is frequently hardly detectable (Zähner 1983). The reason for the relatively large number of known antibiotics and that they have long been known compounds, is the result of their easy detection by simple methods in the screening process. Antibiotics have been isolated so frequently, not because they are the only products, but rather because only those have been screened for. It is very likely that the screening of microorganisms for other types of activity, for example for antiinflammatory activity, will yield numerous new antiinflammatory compounds. It is evident that antibiotics form only a part of the range of bioactive secondary metabolites. In the light of recent findings the term 'antibiotic' is a somewhat outworn concept. In a narrow sense antibiotics are only antimicrobial, antitumour and antiviral natural products, excluding the semi-synthetic derivatives. Today this term must be abolished, or rather extended to all the secondary metabolites which are able to inhibit via a molecular biochemical mechanism various growth processes or replications in prokaryotic or eukaryotic cells at minimal concentration. Compounds filling these requirements may be called 'biological inhibitors'. This definition should cover besides the 'classical antibiotics', insecticides, miticides, antiparasitic compounds (nematocides, acaricides, anthelminthics) herbicides, mycotoxins, cytotoxins, phytotoxins, various growth regulators and also immunity-enhancing compounds (Demain 1983). The borderline between antibiotics and other secondary metabolites is becoming less clear.

1.4 IDENTIFICATION

Although isolation and separation techniques, especially automated chromatographic methods, have in recent years improved to an almost unbelievable degree, the most time- and labour-consuming step of the screening process remains the isolation of pure compounds from the broth. One of the biggest and increasing problems in screening is the earliest possible elimination of unnecessary duplication (Hamill 1977). The second key-step, besides the basic screen, is definitely the early identification of detected metabolites. This is usually performed during the secondary screen, but identification may be possible at different phases of the screening work.

The first step in the identification of an unknown product is to classify it as a chemical type. This can be done by reference to known structural classes having similar properties. If we could identify the active components already in the broth that would be the ideal situation. Unfortunately, except for a very few examples, this is impossible. The criteria for initial characterisation of cultures including culture morphology, antimicrobial spectra, cross-resistance studies, and chromatographic or electrophoretic methods combined with selective bioautography are less precise (Table 1.8). Biological results are frequently masked by other materials present in the broth. Owing to the presence of these possible interfering compounds, the spectrum of biological activity of the whole culture is not necessarily characteristic. However, the use of a series of resistant test organisms may call our attention to a specific group.

Table 1.8 – Methods for identification of metabolites.

Biological methods:	
Taxonomy of producing strains	
Spectrum of biological activities	
Cross-resistance studies	
Chromatographic methods:	
Paper chromatography (R_f spectra)	bioautography
TLC	chemical reagents
HPLC	UV light
Electrophoresis	
Physical-chemical methods:	
Ultraviolet spectroscopy	
Infrared spectroscopy	
NMR spectroscopy	
MS, HPLC-MS, GCMS	
X-ray diffraction	
Chemical degradation studies	
Computerised data bases:	
Physical or chemical data (IR, NMR, chromatography)	
Overall data (Bérdy System)	

Systematic paper or TLC chromatographic studies using a series of different developing solvents and graphical evaluation, comprising the so-called R_f-spectra (summarised chromatograms), usually give several large groups and represent a good guideline for the isolation process (Aszalos & Frost 1975). Chromatographic methods themselves, without reference materials, give only presumptive results. The combination of biological and systematic chromatographic studies with some chemical results, e.g. acid–base character and spectroscopic (ultraviolet and visible), is relatively powerful. If the active compound in the broth has a characteristic UV spectrum (e.g. anthracyclines, polyenes, etc.), it is sometimes possible to identify the chemical class of compound in a crude extract.

Computer-assisted identification techniques based on IR spectra or TLC data are very helpful, especially in large-scale screens (Woodruff 1983; Strauss 1974). The application of GCMS technique (gas chromatography connected to mass spectrometry) is only occasionally possible. The use of HPLC for detection and identification has progressed very rapidly in recent years and may soon be of great value in the identification process. The gentle field desorption MS technique also appears to be very promising for potential early identification. It is especially useful for analysing impure compounds or mixtures of related compounds, which are a frequently occurring phenomenon in the area of antibiotics.

The danger always exists of discarding new useful analogues during screening. Most of the presumptive chromatographic and spectroscopic methods do not differentiate structures having only subtle differences. The difference between gentamicin and sagamicin or daunomycin and adriamycin in the early phase of screening is almost impossible to detect and unfortunately, in similar situations, a new, potentially useful metabolite could escape detection.

Exact identification in most cases is impossible without the isolation of at least semi-pure materials. If the active compound is obtained in a higher degree of purity, a number of spectroscopic methods are available for identification (Mitscher *et al.* 1977). If the compounds belong to a known class they can be correctly identified by chemical degradation or direct comparison.

Powerful physical methods have reduced the time required for structure elucidation from years, to days, and the sample requirement from grams to milligrams. Today any compound isolated in pure form can have its structure determined if there is sufficient interest in the compound.

It is very difficult to separate and identify the minor components which are present in the broth only in trace amounts. Separation and identification of such closely related compounds usually take place only during the final purification step. The practical importance of minor components is not negligible, as is exemplified by the case of cephalosporin C or kanamycin B. The percentage of minor antibiotics among recently published new antibiotics is about 30–35%. In our own studies with a gentamicin C producing new *Micromonospora* strain we could detect more than 40 aminoglycoside components in the broth, and could isolate almost all known gentamicin-type compounds including sisomicin, sagamicin and others, as well as 10 new antibiotics, one of which is depicted in Fig. 1.4 (Bérdy *et al.* 1977).

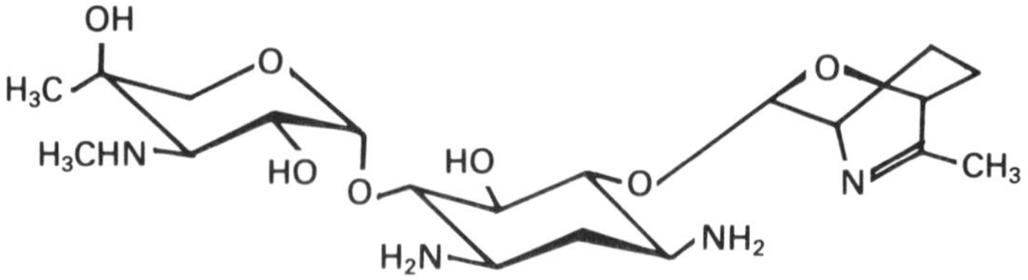

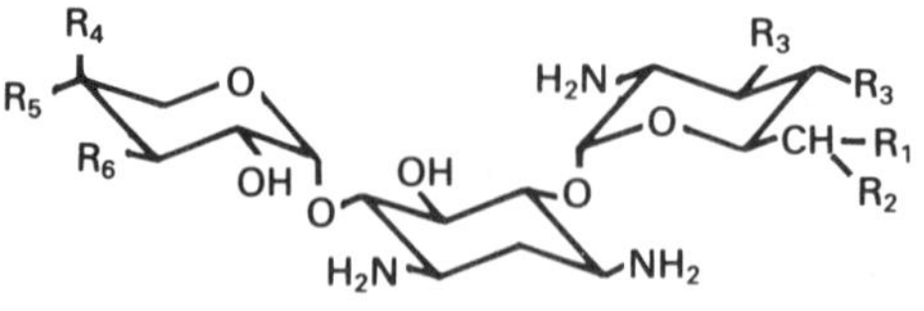

	R_1	R_2	R_3	R_4	R_5	R_6
6′-Methylgentamicin A	CH_3	OH	OH	H	OH	$NHCH_3$
6′-Methylgentamicin A_1	CH_3	OH	OH	OH	H	$NHCH_3$
3″-*N*-Demethylgentamicin C_2	CH_3	NH_2	H	OH	CH_3	NH_2
4″-Demethylgentamicin C_1	CH_3	$NHCH_3$	H	OH	H	$NHCH_3$
4″-Demethylgentamicin C_2	CH_3	NH_2	H	OH	H	$NHCH_3$
4″-Demethylgentamicin C_{1a}	H	NH_2	H	OH	H	$NHCH_3$

Fig. 1.4 – Some aminoglycosides isolated from *Micromonospora* (Bérdy *et al.* 1977).

With some seven thousand known microbial antibiotics and the fact that they are not normally available for direct comparisons, together with the limitations of chromatographic and some spectroscopic identification methods, the importance of various data bases for reference purposes can be seen. Computer-assisted identification based on the comparison of one or more selected characteristics (e.g. IR, TLC, MS) of the unknown compound with the data of known substances, has already proved very useful (Strauss 1974).

The early and correct identification of an isolated crude compound, or the earliest indication of its chemical group by comparison of the overall physical, chemical and microbiological data with those of all known antibiotics today, is indispensable to the identification process. A computerised data bank including the various properties of all known antibiotics, emphasising those that may be characteristic of the crude compounds, strongly facilitates rapid identification. To maintain our current awareness in the continuously growing field of antibiotics, an easily accessible and up-to-date compilation of published data has a primary significance. Our efforts to meet this requirement were achieved in 1975/76 when, with the cooperation of the scientists of the NCI, FCRC† in Frederick, we established a computerised data bank, based on my card index file system (Bostian *et al.* 1977). This compilation covers the data listed in Table 1.9. This system has proved to be particularly useful in practice. In some cases the exact identification of a newly isolated crude compound required only a few days.

Table 1.9 – Characteristics included in the antibiotic data base.

Antibiotic name, synonyms	Chromatography
Chemical type†	Stability
Chemical formula	Test organisms
Elementary analysis	Toxicity data
Antibiotic code number‡	Antitumour/antiviral effects¶
Producing organisms	Isolation methods:
Molecular/equivalent weight	Filtration
Appearance–physical characteristics§	Extraction
Optical rotation	Ion-exchange
Ultraviolet spectra	Absorption
Solubility	Chromatography
Qualitative chemical reactions	Crystallisation

† and ‡ According to the Bérdy classification system.
§ Colour, crystal structure, etc.
¶ *In vitro* and *in vivo* characteristics.

The *Handbook of Antibiotic Compounds* Series mentioned previously is also based on this data bank (Bérdy 1980–82). The continuously (half-yearly) updated complete tape containing the appropriate search and indexing programs is also available.

† National Cancer Institute, Frederick Cancer Research Center.

1.5 EPILOGUE

The number of antibiotic compounds according to my most recent estimate is exactly 10 986, including 7653 microbial metabolites. The number of other known bioactive or inactive microbial metabolites is also increasing very rapidly. The question logically emerges, whether there exists any limit in the increasing number of compounds discovered from microbes in the future? It is a fact that until now only a small proportion of microbes have been screened for their antimicrobial, and a much smaller part for their other bioactivities. Only 20–25% of the bacterial and 10–15% of the known 50 000 fungal genera have been investigated thoroughly.

The production of secondary metabolites in microbes is due to the almost infinite number of genetic units. Obviously the number of potential antibiotic producers and the cultural variations during their fermentation are practically unlimited. Microbial transformations and genetic manipulations represent further increasing factors in the productivity of microbes. According to recent microbial genetics work, it is very likely that in most microorganisms there is a latent possibility for the production of secondary metabolites, but methods and techniques to detect their activity or to increase their production to the normally detectable level have to be developed. In practice only some technical problems and the limited manpower may restrict the expansion of the pool of known microbial metabolites. Systematic and sophisticated evaluation of the microbial world under the most diverse conditions will surely produce an increasing number of new and useful metabolites in the future.

REFERENCES

Aszalos, A. & Frost, D. (1975) *Methods Enzymol.* **43**, 172.

Bérdy, J. (1961) Ph.D. Thesis, Debrecen, in Hungarian.

Bérdy, J. (1970) *Progr. Antimicrob. Anticancer Chemother., Proc. Int. Congr. Chemother., 6th, 1969*, Vol. II, p. 1006.

Bérdy, J. (1972) *Inform. Bull., Intern. Cent. Inform. Antibiot.* **10**, 1.

Bérdy, J. (1974) *Adv. Appl. Microbiol.* **18**, 309.

Bérdy, J., Kádár-Pauncz, J., Méhesfalvi-Vajna, Zs., Horváth, Gy., Gyimesi, J. & Koczka, I. (1977) *J. Antibiot.* **30**, 945.

Bérdy, J. (1980) *Process Biochem.* **15**, 7, 28.

Bérdy, J. (ed.), Aszalos, A., Bostian, M., & McNitt, K. (1980–82) *Handbook of Antibiotic Compounds*, vols. I–X. Boca Raton, CRC Press.

Bérdy, J. (1983) *Proc. 2nd Eur. Conf. Ind. Microbiol., Riv. Biol.* **76**, 347.

Bostian, M., McNitt, K., Aszalos, A., & Bérdy, J. (1977) *J. Antibiot.* **30**, 633.

Demain, A. (1983), *Science* **219**, 709.

Douros, J. D. (1976) *Canc. Trtmt. Rep.* **60**, 1069.

Hamill, R. L. (1977) *Jap. J. Antibiot.* **30**, *Suppl.* S-164.

Hamill, R. L. (1982) In: Bulock, J. D., Nisbet, L. J., & Winstanley, D. J. (eds.). *Bioactive Microbial Products.* London, Academic Press, 71.

Kurylowicz, W., Raczynska-Bojanowska, K., & Kowszyk-Gindifer, Z. (1975) *Post. Hig. I. Med. Dosw.* 385.

Mitscher, L. A. Omoto, S., & Foltz, R. L. (1977) *Jap. J. Antib.* **30**, *Suppl.* S-246.

Ninet, L., Bost, P. E., Bouanchaud, D. H., & Florent, J. (eds.) (1981) *The Future of Antibiotherapy and Antibiotic Research.* London, Academic Press, pp. 375–477.

Round Table Discussion (1970) *Progr. Antimicrob. Anticancer Chemother., Proc. Int. Congr. Chemother., 6th, 1969,* Vol. II, pp. 1102–1103.

Strauss, D. G. (1974) *J. Antibiot.* **27**, 805.

Umezawa, H. (1972) *Enzyme Inhibitors of Microbial Origin.* Tokyo, University of Tokyo Press.

Umezawa, H. (1982) *Ann. Rev. Microbiol.* **36**, 75.

Umezawa, H. (1984) *Canc. Trtmt. Rep.* **68**, 137.

Woodruff, H. B. (1983) *Progr. Ind. Microbiol.* **17**, 71.

Zähner, H. (1978) *Proc. FEMS Symp., Basel 1977.* London, Academic Press.

Zähner, H. (1983) *Proc. 2nd Eur. Conf. Ind. Microbiol., Riv. Biol.* **76**, 359.

2

Approaches to the isolation of an unidentified microbial product

S. J. Box
Beecham Pharmaceuticals Research Division,
Brockham Park, Betchworth, Surrey, RH3 7AJ, UK

2.1 INTRODUCTION

In principle the approaches used in isolation procedures have changed very little over the years. However, recently a number of factors have influenced the detailed aspects. Very large numbers of secondary metabolites are now known and one of the major tasks is to determine at the earliest possible stage whether the compound under investigation is one of these previously described metabolites. To achieve this it is necessary to have available some form of data base, such as that prepared by Bérdy (Bérdy 1980), with which to compare the properties of the unknown. This information is important since it will either aid the extraction studies, or deter further studies. Also, because of the large numbers of compounds already known, the importance of structure determination has in many cases superseded the need to investigate the biological properties of the metabolite. Therefore many isolation procedures are aimed at producing sufficient material to satisfy the chemists dealing with structure determination. With the advent of sophisticated infrared and nuclear magnetic resonance spectrophotometers and mass spectrometers, often the quantities required for full structure determination are very small so extraction processes can be carried out on smaller volumes of culture filtrate, making the whole problem easier. To some extent this advantage has been offset by the use of more sensitive screening procedures. For example, one of the groups of compounds described later was originally detected in the fermentation at a level of approximately 0.1 mg dm^{-3}. Before moving on to some topics in more detail, it is important to highlight the significance of high performance liquid chromatography in extraction studies. Its sensitivity and selectivity has made it of critical importance as an assay procedure. In addition, it has proved extremely valuable in a semi-preparative mode where its ability to achieve very good resolution, very rapidly has made it particularly useful when handling unstable compounds.

2.2 TECHNIQUES

2.2.1 Detection methods

The first requirement is the availability of an adequate assay system. Screening proceedures may be developed which can detect an 'effect' in culture filtrate samples but it is not always possible to translate this test into an adequate assay procedure. This indeed may be a reason why, although many antibacterially active metabolites are known, few have been reported with pharmacological activity since it is often difficult to translate the tests used for the latter type of activity into an assay system suitable for use in isolation studies.

There are four signficant factors which should be considered in relation to the development of an assay system for isolation studies. The first is quantification. Although this is implicit in the word assay, it is important that the original screening procedure is adaptable to give such quantitative results. Extraction procedures do not normally work absolutely and it is therefore essential to have some means of assessing accurately the value of different approaches. The assay procedure must also be fairly rapid: If one is reliant on a biological assay procedure which takes, say, five days, then unless the test compound is very stable it will have degraded before there is time to process it further. Even if it is stable progress will be extremely slow. The third important factor is high throughput. Extraction processes generate increasingly large numbers of samples, so an assay procedure which can be completed in three hours but which can accommodate only one sample is really no use. It is probably necessary to accommodate a minimum of say 20 samples per day. Perhaps the most important requirement of the assay is selectivity since the required metabolite will almost certainly represent less than 1% of the dry weight in the culture filtrate. Biologically based test systems are normally sufficiently selective. Alternatively TLC using spray reagents or HPLC can be used. However, the more complex the procedure the slower the throughput and often the poorer the accuracy.

2.2.2 Preliminary investigations

Assuming that an adequate assay procedure and a culture capable of producing the required compound are available, what procedures can be used to identify a suitable extraction?

Given a successful fermentation, the first factor to investigate is whether the activity is associated with the cells or resides mainly in the culture filtrate. If the activity is associated with the cells then the first task is to release the activity. This can be achieved in a number of ways. Simple cell disruption using techinques such as ultrasonication, treatment of the cells with solvents such as acetone, or in some cases treatment of the cells with hot water have all proved successful. After removal of the cell debris and any organic solvent, the resulting aqueous phase can be treated in much the same way as culture filtrate.

The following preliminary tests applied to culture filtrate or cell extracts should give some indication of potentially useful isolation procedures. The first parameter to measure is stability at a range of pH values and temperatures. As a guide, pH values of 3, 5, 7 and 9 could be used, with temperatures ranging from 4°C to 100°C. The period of such tests can be varied depending on the extremes of the test conditions. The next useful information to obtain is on the solvent extractability of the compound. Here it is valuable to use a number of solvents of differing polarities (e.g. butanol, ethyl acetate, chloroform) at a range of pH values. A problem here is that the assay procedure may not be useful for testing samples in organic solvents. If the solvent is removed by evaporation the active compound may be unstable, or alternatively it may be difficult to resuspend a non-polar compound in water, so care must be taken with the design of such experiments.

Determination of ionic character can be very valuable and is a matter of answering the question: 'is it acidic, basic, neutral or amphoteric?' Electrophoresis is a most valuable technique, and determination of the mobility of compounds at a range of pH values can provide critical information. Once again there may be problems in assaying the results of such experiments. A possible alternative is to test the adsorption of the activity to a range of ion exchange resins in 'mini' columns.

The final property to highlight here is molecular size. At this stage the tests should be fairly limited and restricted to differentiating between 'truly' high molecular weight compounds such as proteins and polysaccharides and low molecular weight compounds using techniques such as dialysis.

The significance of this series of experiments is not only in their individual results but also on building an overall view of the properties of the compound. The results of one test will have a bearing on the interpretation of the results of another and help to build this overall picture. Although it may be possible to think of alternative properties which could prove valuable, those highlighted above should provide an adequate basis for starting such isolation studies.

2.2.3 Purification procedures

Having determined some preliminary properties for the target compound what are the basic techniques that can be used to purify that component? Before identifying these techniques it is important to recognise the magnitude of the problem. In many cases the active component will represent less than 0.5% of the total dry weight of the culture filtrate. It is also important to recognise that all possible extraction techniques do not have the same value at different stages in the process. For example, the major requirement of the first stage is that it can be operated in the presence of large volumes of water containing a significant quantity of suspended solids, whereas later stages involve high resolution of small quantities of material.

The general approaches that may be used are as follows.

2.2.3.1 *Solvent extraction*

This is probably one of the most attractive first steps in microbial metabolite isolation studies. The use of a water immiscible solvent permits rapid removal of water, the resulting extract usually in a reduced volume being a more easily handleable material. In terms of the suitability of a solvent, it is best to use the most selective (probably least polar) solvent, even if in some cases this means incomplete extraction of the target compound. In the case of genuinely lipophilic compounds, such extractions are generally most effective at pH values near neutrality. Compounds containing a simple acidic or basic function can be rendered solvent soluble in some cases by adjusting the pH to that at which it is non-ionic. With these types of compounds a further purification can be achieved by back extraction into water at a pH at which the functional group is ionised. Solvent extraction techniques can be modified using methods such as ion pairing. The use of such an approach will be exemplified later in this chapter.

2.2.3.2 *Ion exchange*

Procedures based on ion exchange techniques probably offer the widest range of choices and can be used either at an early stage in the extraction or chromatographically for more difficult separations at later stages in the purification process.

In general ion exchange materials are produced by chemically binding acidic or basic functional groups to an inert support. When culture filtrates containing acidic metabolites are passed down an anion exchange column (basic functional group), these metabolites are retained, whilst basic and neutral metabolites pass through. The retained acids can be eluted from the resin by exchange with higher affinity ionic groups. The same principle applies for basic metabolites and cation exchange resins.

The different ion exchange products available can be grouped by the nature of the support matrix. The most common group is that based on the synthetic polymeric beaded resins which are either polystyrene or acrylic based. The chains of polystyrene are crosslinked by divinyl benzene (DVB) to form an insoluble porous lattice onto which are bonded the ionic groupings. The properties of the resin are affected by the degree of crosslinking of the styrene polymer with the divinyl benzene, and are normally available with crosslinking in the range 2% to 12% DVB. A resin with a low crosslinking has a more open structure and is permeable to higher molecular weight substances, and has a tendency to swell and shrink more but equilibrates more rapidly. Most resins are also available in a range of bead sizes. In addition to the variations of the polymer matrix these resins are available with a range of ion exchange groups including strongly and weakly acidic and basic groupings. For those unfamiliar with such resins, many manufacturers of ion exchange resins produce technical manuals which can provide a useful starting point for further studies. In terms of applications of particular resins, large beaded, low crosslinked resins are probably most suitable for primary extraction procedures from culture filtrates whilst higher crosslinked, smaller beaded resins are more suitable for chromatographic applications.

Another group of ion exchange materials are based on a cellulose support to which the functional groups are bound. These supports are generally not suitable for processing crude culture filtrates and are in fact recommended for the separation of biopolymers although they have found use in the chromatograhic separation of a wide range of secondary metabolites. A further group of exchangers, the ion exchange gels, are available and are based on either a dextran polymer (Sephadex series)† or agarose polymer (BioRad series), their uses are much the same as outlined for the cellulose group. All these supports are available in strong and weak acidic and basic forms.

Ion exchange separations can be applied in two ways. The batchwise approach, normally used at an early stage in an isolation procedure, where the products are adsorbed and eluted with no attempt at a chromatogrphic separation on the column. Second, at later stages a chromatographic approach can be used in which columns are eluted with salt or pH gradients to separate components on the column.

2.2.3.3 *Adsorption techniques*

The use of surface adsorption is a very valuable technique in microbial metabolite isolation, both in a batchwise and chromatographic mode. The common requirement amongst these agents is a high surface area, much of which is contained in pores which restricts their use to lower molecular weight compounds (mol. wt. <2000). In practice the supports are most useful for adsorbing non-polar compounds from polar solutions, the adsorbed materials being eluted with less polar organic solvents.

The most common adsorbents are silica, alumina, carbon and, more recently, a range of polystyrene-based polymeric adsorbents. Carbon in its various forms has probably been the most widely used of these materials. the properties of the carbon vary depending on its source and it can be used at stages throughout an isolation procedure. It finds its greatest applicability in the purification of non-polar compounds.

Perhaps the most interesting members of this group are the polymeric resins using essentially the unfunctionalised matrix of the ion exchange resins (divinyl benzene crosslinked polystyrene). These products have the advantage of being better defined than the carbons and can carry out similar processes. In large bead form products such as Amberlite XAD-4 and Diaion HP20 can provide excellent first steps in an extraction process providing high throughput and good capacity. In small bead form the resins give excellent chromatographic separations.

2.2.3.4 *Gel filtration*

Separations using these supports are based on molecular size. Examples of supports

† See end of chapter for suppliers of proprietary materials.

are crosslinked dextrans (the Sephadex range) or crosslinked polyacrylamide (the Biogel range). The resins are supplied as beads permeated by pores. With increasing crosslinkage the matrix has smaller pores which allow the entry of only lower molecular weight compounds, whilst with less crosslinking larger pores are formed and larger compounds may enter. By careful control of crosslinking a range of matrices are available which can separate compounds with molecular weights within the nominal range 700 to 800 000. In general separations are restricted to water-soluble compounds, although a limited number of products are available for use with organic solvents. As with the ion exchange resins, the manufacturers supply useful literature as a starting point for studies with these supports.

In general these separations are restricted to the later stages of a process and in some cases may be used to desalt products following ion exchange processes.

2.2.3.5 *Partition chromatography*

Partition chromatography using supports such as cellulose and silica gel is restricted to the latter stages of the isolation procedure and is generally used to complete the more difficult separations.

The techniques highlighted above are not intended as an exhaustive survey. There are other more selective techniques, such as affinity chromatography, which can be used. However, the methods which have been mentioned will provide a valuable starting point for the isolation of most microbial products.

One important aspect to remember in any extraction process is that although each step removes impurities it also concentrates compounds of similar properties so that separations become increasingly difficult and demanding. In general it is probably easy to remove the first 95% of impurities but difficult to remove the remainder. The chances of obtaining a complete purification will always be enhanced by utilising a range of processes based on different properties of the target compound.

2.3 APPLICATIONS

A number of extraction processes are outlined below and these will exemplify the use of the techniques outlined above.

2.3.1 Prasinons

The first isolation procedure relates to two insecticidally active metabolites, the prasinons (Box 1973) which were isolated from a culture of *Streptomyces prasinus*. Before discussing the extraction there are two ancillary points of interest. First the assay procedure used for these compounds was based on activity against the sheep blow fly larva (*Lucilia sericata*). This test presented two problems, the samples required a particular preparation procedure so that they were suitable for testing and the test itself required some days for completion. This meant that progress with the extraction procedure was limited by the length of the assay and the preparation of test samples and it was not until the activity was correlated with discrete zones on a TLC system that rapid progress was made. This highlights the importance of a suitable assay procedure for extraction studies.

The second point relates to the selection of the starting material. It may seem that it is only necessary to culture the producing organism under a number of conditions and select the filtrate with the greatest activity. The results in Table 2.1 show this may not always be true. The prasinons are solvent extractable compounds and ethyl acetate was the sovlent of choice. If the weight of extract and activity for each of the two media are compared there is a clear choice between a larger amount of less pure target compound in medium B as against less but purer prasinon in medium A. These differences are often manifested in the case of *Streptomyces* cultures when cultures are grown in synthetic and complex media. Careful consideration of each case is required before the correct selection can be made.

Table 2.1 – Insecticidal activity of extracts from *Streptomyces prasinus* fermentations.

Medium	Extract from culture filtrate	Larvicidal activity of extract†		
		50 μg cm^{-3}	10 μg cm^{-3}	2.5 μg cm^{-3}
A	6‡	100	100	69
B	75‡	100	88	0

† % kill.

‡ weight of extract obtained from 1 millilitre culture filtrate.

The procedure that was developed for the extraction and purification of the prasinons is shown in Fig. 2.1. This procedure is typical for the purification of such lipophilic compounds. It exemplifies how useful a solvent extraction step can be; this stage reduced the volume from 200 litres of culture filtrate to 50 litres of isobutylacetate which was readily removed by evaporation. In addition it was possible to estimate the purification in terms of activity on a dry weight basis as some 120-fold. There is nothing very remarkable about the remainder of the processing, which is mainly based on silica gel chromatography. However, the process does include the use of Sephadex LH-20. This support was developed to undertake molecular size separations in organic solvents, and, although nominally relying on this molecular size differentiation, much of the separation in this case also probably relied on adsorptive interactions with the matrix. This work was carried out over 10 years ago and although significant quantities of the metabolite were prepared their structures were never completely defined. It is possible that today, with the availability of improved analytical equipment, greater success in this aspect would have been achieved with less material.

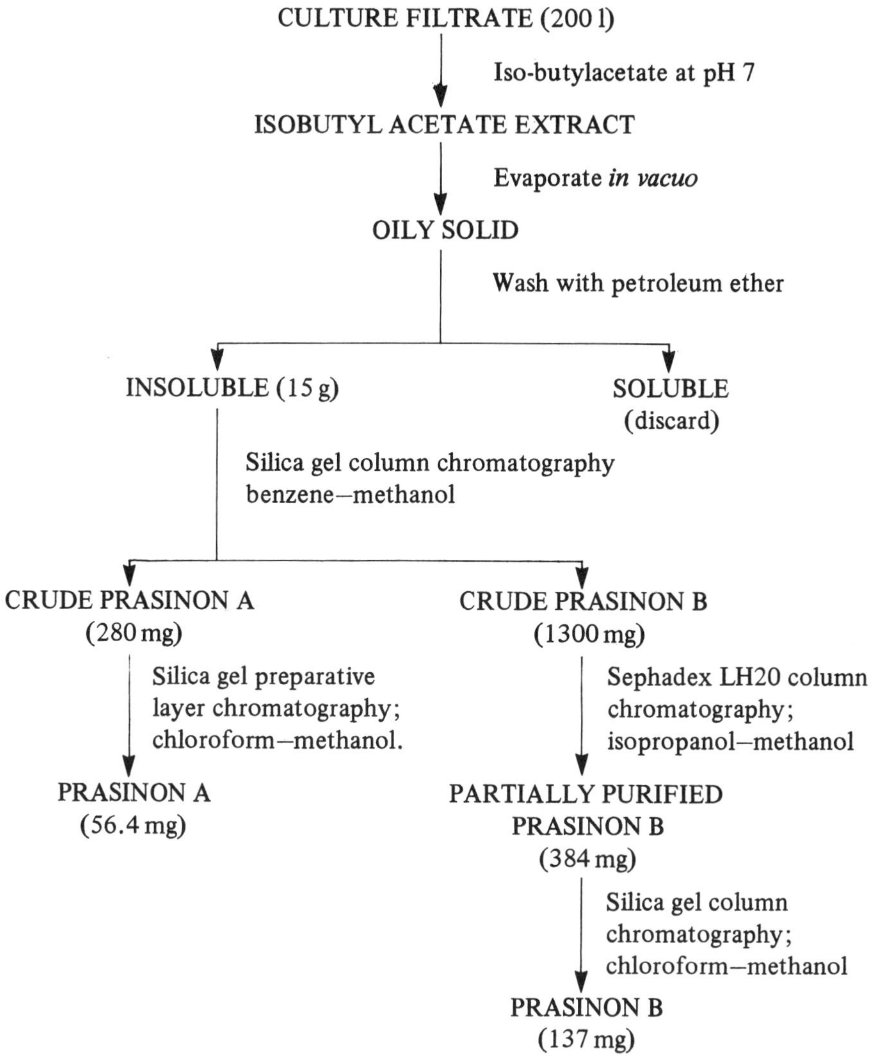

Fig. 2.1 – Isolation procedure for Prasinons A and B (Box *et al.* 1973). (Reproduced by permission of the American Society of Microbiology.)

2.3.2 β-Lactams

The next group of isolation procedures which will be highlighted relate to members of one of the largest and most commercially significant groups of antibiotics, the β-lactams. Although the major groups, the penicillins and cephalosporins have been known for a long time a resurgence of interest in screening led to the discovery of a number of new types in the last decade. These groups include the cephamycins (Miller *et al.* 1972), nocardicins, clavulanic acid, the carbapenems (Brown 1981; Parker *et al.* 1982), and the monobactams (Sykes *et al.* 1981).

2.3.2.1 *Clavulanic acid*

The first extraction process to be dicussed is that of clavulanic acid. The structure of clavulanic acid is shown in Fig. 2.2. Clavulanic acid is not used clinically as an antibiotic but it inhibits the bacterial enzyme β-lactamase which confers resistance on certain clinical strains to β-lactam antibiotics.

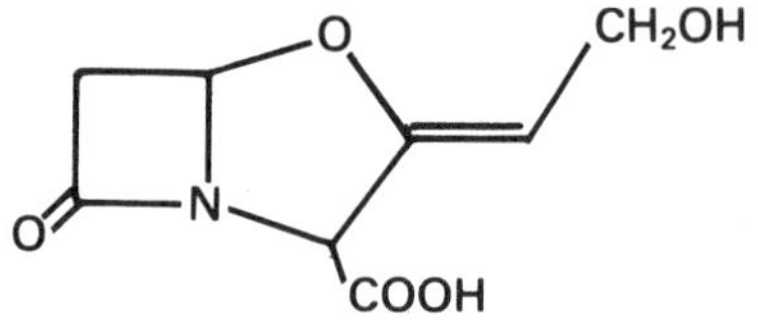

Fig. 2.2 – The structure of clavulanic acid.

When this compound was first detected, some investigations of the paper chromatographic properties of the compound were undertaken. The R_f values obtained in neutral and acidic solvent systems were 0.13 and 0.66 respectively, suggesting the active component had a higher affinity for the organic mobile phase at acidic pH than neutral. In fact clavulanic acid was extracted into butanol at pH 2.0 and this step formed the basis of the procedure outlined in Fig. 2.3 (Reading and Cole 1977). Clavulanic acid is only poorly stable at pH 2.0 and it was necessary to carry out the extraction rapidly and in the cold to improve recoveries. Additional indications from the paper chromatography suggested that clavulanic acid would back extract into water at pH 7.0 and this indeed proved to be the case. The back extraction not only had the advantage of being a valuable purification but also returned it to a condition under which it was more stable. Now that the structure is known it is interesting to note that this is an example of a normally hydrophilic compound being rendered solvent soluble by the appropriate pH adjustment. The remaining stages are straightforward. Once it was recognised that the new compound was acidic, chromatography on an ion exchange resin was a logical procedure and purification was improved by the use of a sodium chloride gradient elution. Desalting was achieved using molecular size separation on a gel column, which probably also involved some separation of organic impurities. The final purification was achieved by partition chromatography on cellulose.

Before leaving clavulanic acid it is interesting to compare one of the alternative procedures developed for the purification of this compound which is summarised in Fig. 2.4 (British Patent 1,563,103). The preliminary steps are essentially those used to isolate clavulanic acid as its salt. The effect of applying isolation steps is not only to separate some contaminants but also to concentrate compounds with like properties, so that the separations often become more difficult as the process proceeds. A way to overcome this is to apply at each stage a separation based on a different character of the compound. An alternative is to use the approach demonstrated here. In this case the compound was derivatised during extraction. Esterification of the acid function changed the characteristics of the clavulanic acid allowing the application of a different range of separative procedures and effective purification of the ester. If the extraction procedure is aimed only at structure determination this derivatisation need not be reversible, but if the final product is required for biological evaluation then the ability to regenerate the starting material is essential. Of course, before derivatisation can be used, an adequate assay for the derivative must be available, since it may not retain its biological activity in this form. Derivatisation may also be used in the case where an unstable compound can be stabilised to improve extraction efficiency.

Culture supernatant

↓ HCl to pH 2.0; 3/4 vol. of butan-l-ol
Extraction at 5°C

Butanol extract

↓ 1/15 vol. of water to pH 7.0 with 20% (wt/vol.)
aqueous NaOH

Aqueous back extract

↓ Ion-exchange chromatography;
permutit FFIP (SRA 62) resin Cl-, NaCl 0–0.35 M
gradient elution, 5°C; combined active fractions
vacuum concentrated

Concentrated active eluate

↓ Desalted on BioRad Bio-Gel P2 column at 5°C
and vacuum concentrated combined, active,
NaCl-free fractions

Concentrated desalted material

↓ Column chromatography at 5°C on Whatman CC31
cellulose using butan-l-ol–ethanol–water,
4:1:5 (top phase); active fractions combined,
evaporated to dryness, dissolved in distilled water
and freeze-dried

Sodium clavulanate

Fig. 2.3 – Isolation procedure for sodium clavulanate (Reading & Cole 1977).
(Reproduced by permission of the American Society of Microbiology.)

Culture filtrate

↓ Adsorb onto Zerolit FF-IP SRA 61, (anion exchange resin), elute with 1 M NaCl

Combine fractions containing clavulanic acid

↓ Chromatograph on Amberlite XAD-4 elute with deionised water

Combine fractions containing clavulanic acid

↓ Concentrate by reverse osmosis and freeze-dry

Freeze-dried solid

↓ Treat with benzyl bromide in dimethylformamide

Impure benzyl clavulanate

↓ Chromatograph on Sephadex LH20 elute with cyclohexane–chloroform (1/1)

Combine fractions containing benzyl clavulanate and concentrate by evaporation

↓ Chromatograph on silica gel elute with cyclohexane/ethyl acetate (1/1)

Benzyl clavulanate

(Clavulanic acid can be regenerated by hydrogenolysis)

Fig. 2.4 – Purification of clavulanic acid benzyl ester.

2.3.2.2 *Carbapenems*

Another group of β-lactam anitbiotics which have recently been isolated are collectively known as the carbapenems. These compounds all contain a fused β-lactam system; however, one significant difference between these and clavulanic acid is that some 40 related compounds have been isolated from various *Streptomyces* species. Although all 40 are not known to occur in any single culture, in many cases they produce 10 or more closely related compounds. This of course presented significant difficulties as far as assaying was concerned and was an example where HPLC made a very significant contribution.

The process reported below relates to the extraction of three of the earliest members of this group collectively known as the olivanic acids and individually designated MM 4550, MM 13902, and MM 17880 (Butterworth *et al.* 1979; Hood *et al.* 1979). The structures of these compounds are shown in Fig. 2.5. The isolation of these metabolites was extremely difficult and continued over a significant period. At the outset only MM 4550 was detected. Early attempts at isolation proved fruitless; this failure was thought to be due to instability. However, much later, when the compounds were finally purified, it was realised that the original titres were less than 0.1 mg dm^{-3} MM 4550 (Hood 1982). So, although instability undoubtedly played a part, the extremely low levels were probably most responsible for 'losing the compound' in these early studies.

MM 4550

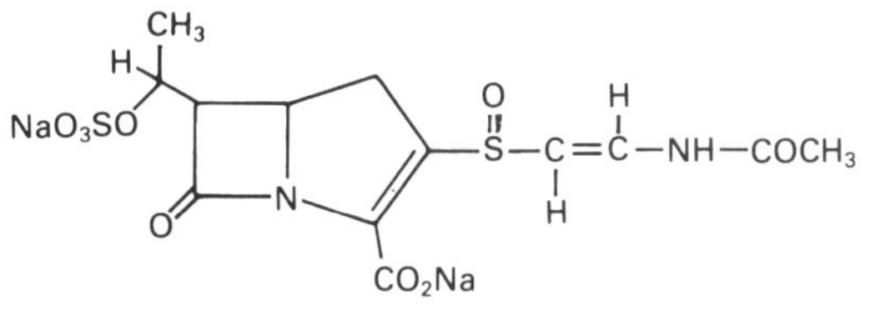

MM 13902

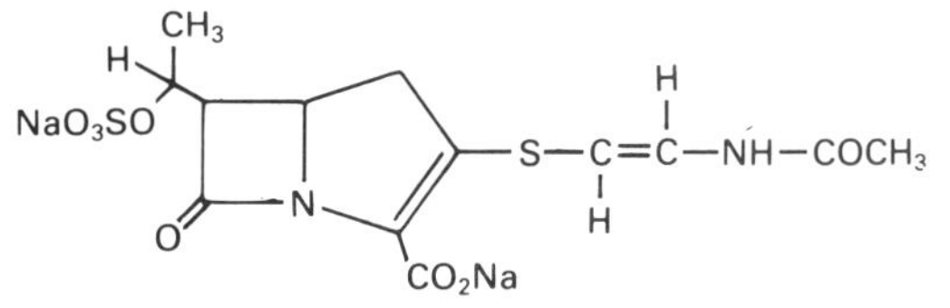

MM 17880

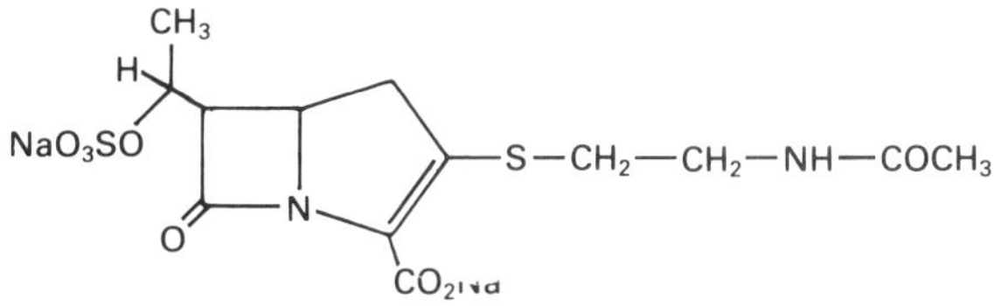

Fig. 2.5 – Structures of MM 4550, MM 13902 and MM 17880.

MM 4550 was shown to occur almost totally in the culture filtrate and electrophoretic studies suggested it was a diacid. Initially ion exchange techniques using standard resins were tried, but all activity was lost; again this was assumed to be due to poor stability. The acid instability of the MM 4550 precluded direct solvent extraction when the pH was adjusted to a level where protonation would occur. However, the use of ion pair extraction provided an important first step. In this technique a lipophilic quaternary ammonium salt is dissolved in a non-polar solvent, which is used to extract the culture filtrate. The strongly acidic antibiotics are extracted into the organic phase to form a hydrophobic complex with the quaternary ammonium salt. Efficiency of the extraction is dependent on the lipophilicity of the quaternary ammonium salt as is demonstrated by the results in Table 2.2. Thus tetra-*n*-butylammonium chloride gave little or no extraction whilst the longer carbon chain lengths give steadily increasing efficiency. Obviously selection of a particular system is based on limited information during early studies and optimisation will follow during scale up. Once the ion pair is generated it can be purified in this form or alternatively it can be back-extracted into the aqueous phase by regeneration of the salt. In this case back-extraction was carried out using iodide as the counterion. Using benzyldimethyl-*n*-hexadecyl ammonium chloride as the quaternary ammonium salt and sodium iodide as the counterion this overall stage gave a volume reduction from culture filtrate of greater than 200-fold with an approximately 40-fold purification. This stage formed the basis for the procedure outlined in Fig. 2.6.

Table 2.2 – Comparison of quaternary ammonium salts as ion pair extractants for the olivanic acids (Hood *et al.* 1979).

Quaternary ammonium salt	Extraction efficiency (%)
Benzyldimethyl-*n*-hexadecylammonium chloride	65
Trioctylmethylammonium chloride	73
Tetra-*n*-butylammonium chloride	Trace
Dimethyldioctylammonium chloride	41
Dimethyldidecylammonium chloride	73

Reproduced by permission of the *Journal of Antibiotics*.

Looking at some further stages in the procedure it is interesting that the three components MM 4550, MM 13902 and MM 17880 are separated by ion exchange chromatography on the strongly basic ion exchange gel, QAE Sephadex. Since the ionic properties of the compounds are so similar such a separation might not have been anticipated. Indeed, as demonstrated below, the separation probably owed more to secondary adsorption to the gel than to ion exchange effects (Butterworth *et al.* 1983).

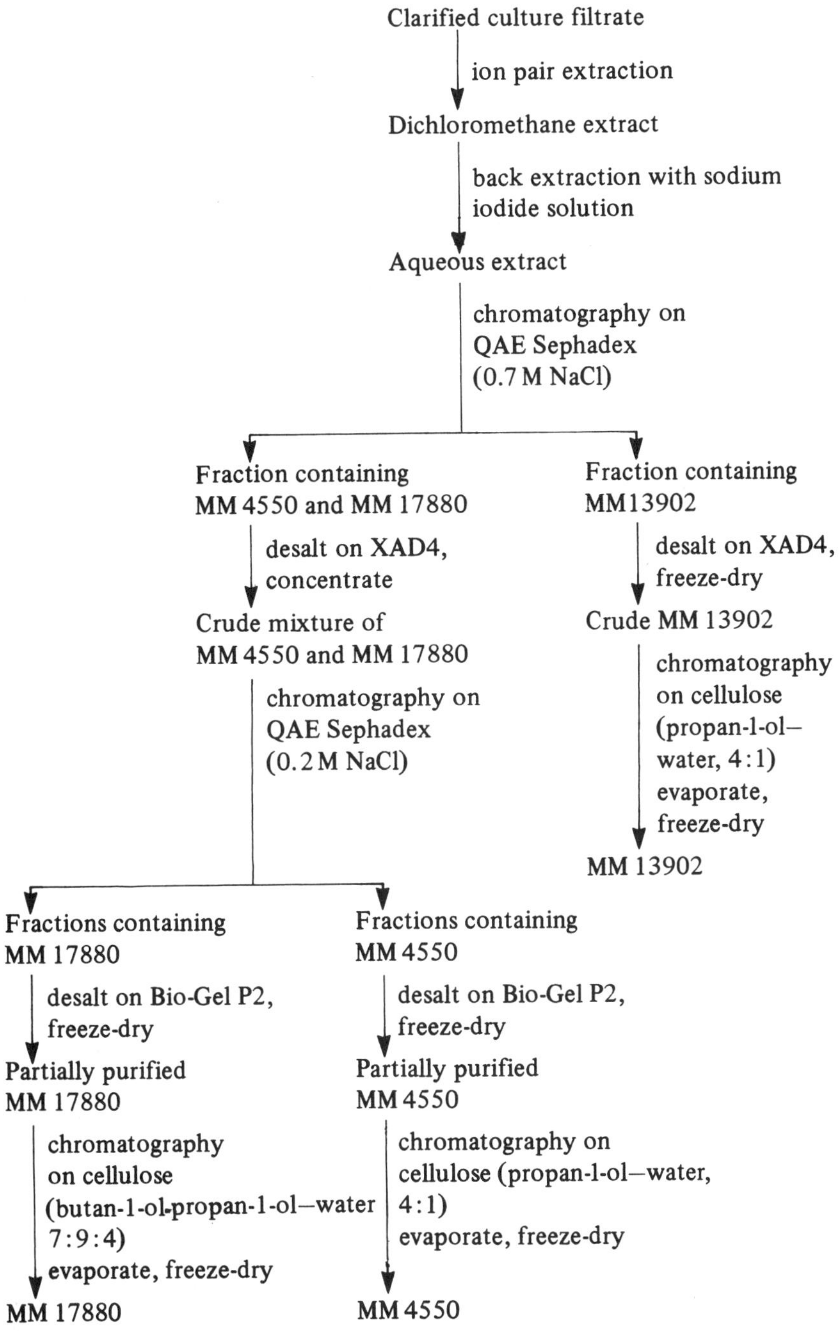

Fig. 2.6 – Isolation procedure for MM 4550, MM 13902 and MM 17880 (Hood *et al.* 1979). (Reproduced by permission of the *Journal of Antibiotics.*)

The results in Fig. 2.7 demonstrate the contribution of secondary hydrophobic adsorption in the separation of MM 13902 and MM 17880 on QAE Sephadex. The presence of 3.2% butanol largely overcomes the hydrophobic binding and the adsorption isotherms reflect mainly ionic interactions and at increasing salt concentrations the compounds display similar affinities for the resin. However, in the absence of butanol, the affinity differences increase with increasing sodium chloride concentration, an effect characteristic of hydrophobic adsorption to the gel.

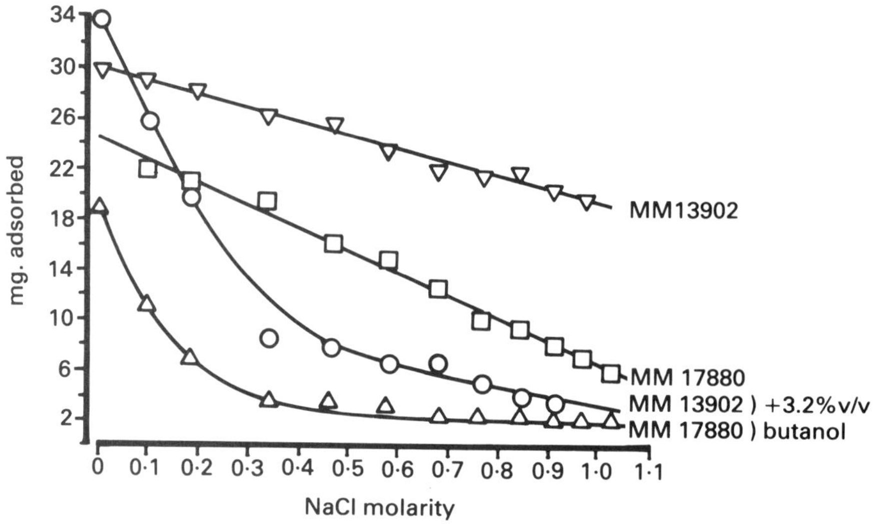

Fig. 2.7 – Adsorption of MM 13902 and MM 17880 onto QAE Sephadex A25 (1 g adsorbent used in all cases), (Butterworth *et al.* 1983). (Reproduced by permission of Alan R. LISS Inc.)

The desalting stage using Amberlite XAD-4 macro-reticular non-ionic resin, having the basic structure of that found in many ion exchange resins, is also of interest. It was found that in the presence of sodium chloride the olivanic acids were adsorbed onto the resin, but in the absence of salt could be eluted with distilled water. This procedure, somewhat akin to salting out in solvent extraction, was valuable in desalting the olivanic acids although ironically it was often necessary to add additional salt to get complete binding to the resin. Since the polystyrene divinyl benzene matrix of XAD-4 also formed the basis for the majority of ion exchange resins this pointed to the reason for the failure of the polystyrene ion exchange resin approaches with these compounds.

The results shown in Fig. 2.8 (Butterworth *et al.* 1983) clearly demonstrate the reason for this failure. When an acrylic-based anion exhange resin (Amberlite IRA 458) is used, it shows, as would be expected for an ion exchange process, decreasing affinity between MM 13902 and the resin at increasing sodium chloride concentrations. In contrast the styrene divinyl benzene resin containing no ion exchange groups (Diaion HP20), which is very similar to XAD-4, displays increasing affinity for MM 13902 with increasing salt concentrations. When the anion exchange resin IRA 400 with a styrene divinyl benzene matrix is used there is competition between ionic and hydrophobic interactions and it is clear that elution of MM 13902 from such a resin will always be poor. Thus hydrophobic interaction with the matrix of the ion exchange resins was to blame for the failure of this approach at early stages in the isolation rather than instability of the olivanic acids.

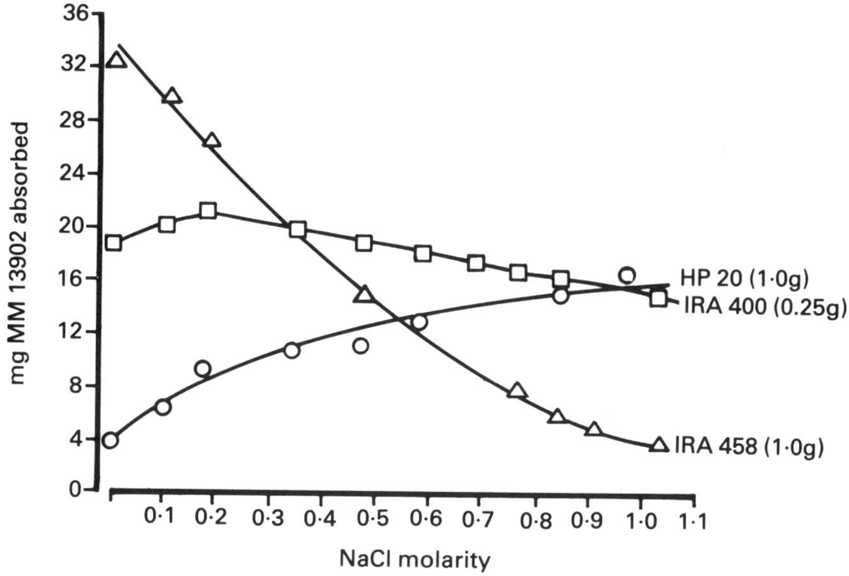

Fig. 2.8 – Adsorption of MM 13902 onto resins (Butterworth *et al.* 1983). (Reproduced by permission of Alan R. LISS Inc.)

Of course these results are looked at with hindsight as far as the purification of MM 4550, MM 13902 and MM 17880 are concerned. But they should serve to demonstrate how early results, when structures are not known, can mislead the extraction scientist's ideas on the properties of the particular component under investigation.

Before leaving this group, Table 2.3 shows the results of stability tests which were carried out on the three components once they were purified. What is interesting is the variation in optimum pH for the different members of the group. It is an indicator that, if, for example, MM 4550 were the predominant component with a stability optimum at about pH 6.0 and process solutions were maintained at this value, this could seriously limit the extraction of the other components. This is a further demonstration of how carefully preliminary results must be evaluated.

Table 2.3 – Stability of olivanic acids in aqueous solution at 25°C (Hood *et al.* 1979).

	$t_{1/2}$		
pH	MM 4550	MM 13902	MM 17880
2.2	13.5 min	2.5 min	5 min
4.0	5.5 h	7.5 min	15 min
6.0	18 h	4 h	5.5 h
7.0	10.5 h	12 h	15 h
7.8	8.5 h	27 h	27.5 h
10.2	19 min	4 h	2.5 h

Reproduced by permission of the *Journal of Antibiotics.*

2.3.2.3 *Epoxyquinone antibiotic*

As a final example the results obtained with a further antibiotic designated MM 14201 are reported below. When preliminary studies were undertaken some years ago little progress was made, owing to the instability of the compound. However, more recent studies have led to the structure of the compound being determined (Box *et al.* 1983).

Studies showed that MM 14201 was very labile with greatest stability around pH 7.0. The only other useful information obtained at this stage was that the compound moved towards the cathode when submitted to paper electrophoresis. All attempts to use solvent extraction, ion exchange and adsorption chromatography failed owing to loss of the MM 14201. Eventually a 'reverse' extraction procedure was used in an attempt to remove impurities and it was during these studies that the biological activity was correlated with a component on a high performance liquid chromatogram. This correlation was critical from two points of view, the rapidity of the HPLC assay was important with such an unstable compound and the HPLC acted as a semi-preparative final stage in the isolation procedure outlined in Fig. 2.9.

The procedure was aimed at the removal of impurities and initially the final HPLC was carried out using a pH 7.0 phosphate buffer as eluant. However, it was noticed that the final preparations were even more unstable and it was realised that the presence of the phosphate buffer which had been used throughout the procedure to stabilise pH was causing greatly increased rates of degradation. Again this indicates how carefully preliminary studies should be undertaken, particularly in defining the conditions for greatest stability. Even in the absence of phosphate, MM 14201 was too unstable to isolate in solid form and a chemical derivatisation was used to finally isolate the acetyl derivative (MT 35214). The structure was finally determined for the compound and is shown in Fig. 2.10. This procedure highlights the usefulness of HPLC as an assay procedure, and emphasises the need to consider chemical derivatisation to aid extraction.

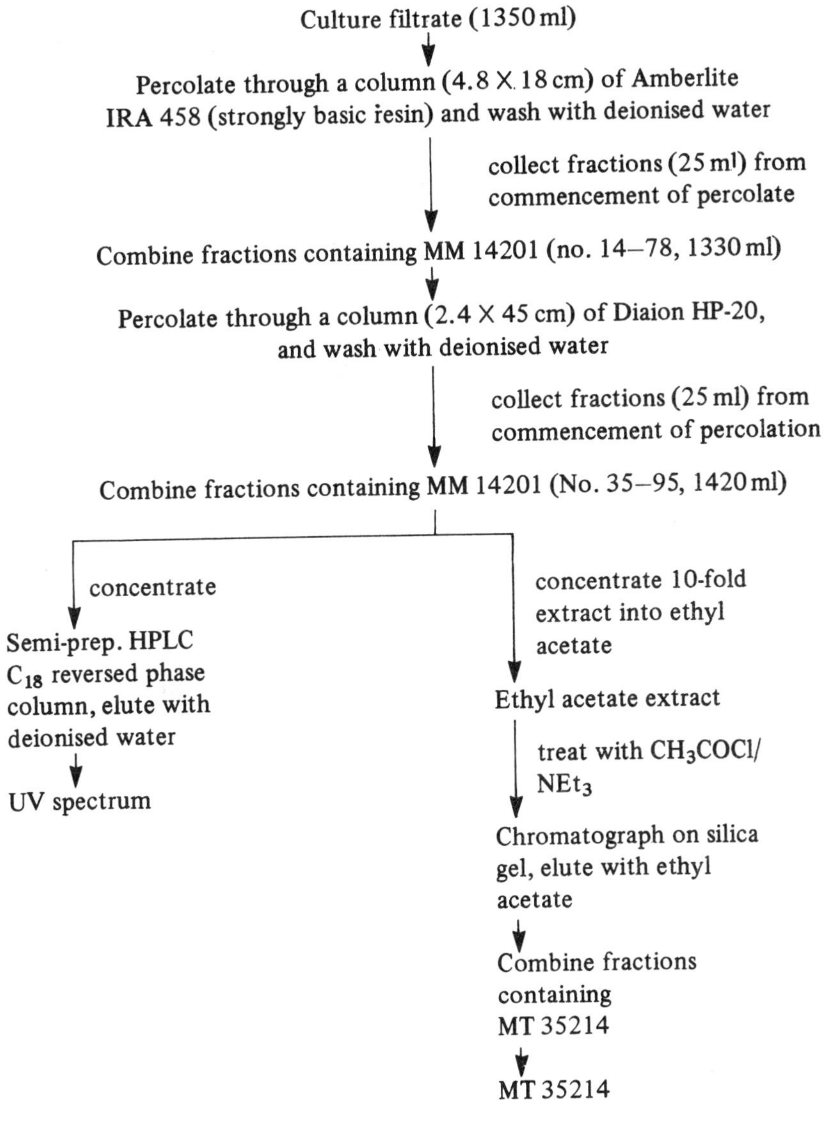

Fig. 2.9 – Purification procedure for MM 14201 and MT 35214 (Box *et al.* 1983). (Reproduced by permission of the *Journal of Antibiotics*.)

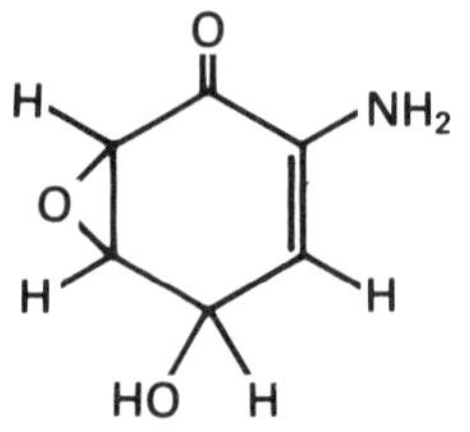

MM 14201

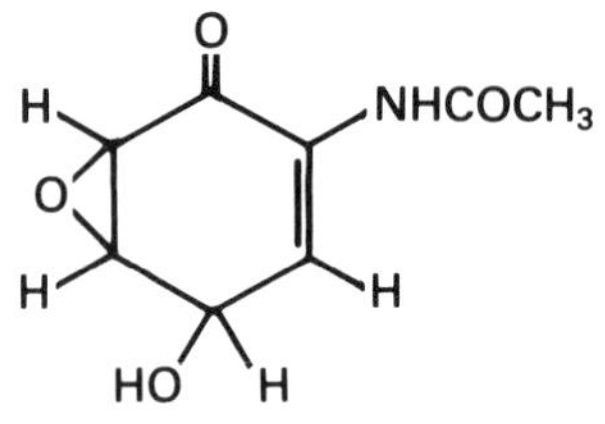

MT 35214

Fig. 2.10 – Structures of MM 14201 and MT 35214.

2.4 SUMMARY

These examples have been intended to demonstrate some significant aspects in the approaches to isolating microbial metabolites, and the following points should be borne in mind at the start of any such study. First, preliminary studies should be carried out most carefully since it is these that set the basis for all future studies. The development of a rapid and selective assay procedure is obviously essential and HPLC is playing an increasingly important role here. Initially it should not be assumed that any particular column support is acting solely by the means for which it was primarily prepared. Finally when dealing with unstable compounds the possibility of using a chemical derivatisation stage should always be considered.

With the introduction of new and more sensitive assay procedures, and improved analytical techniques I am sure there remain many more groups of useful and interesting metabolites to be discovered.

SUPPLIERS OF PRODUCTS USED IN ISOLATION STUDIES

Amberlite resins	Rohm & Haas Company, Philadelphia, USA.
Bio-Gel	BioRad Laboratories, Richmond, California, USA
Diaion resins	Mitsubishi Chemical Industries, Tokyo, Japan
Sephadex	Pharmacia Fine Chemicals, Uppsala, Sweden
Whatman Cellulose CC31	Whatman Ltd, Maidstone, Kent, UK.

REFERENCES

Beecham Group Ltd. British Patent 1,563,103.

Bérdy, J. (1980) *CRC Handbook of Antibiotic Compounds* in 10 volumes. Boca Raton, CRC Press.

Box, S. J., Cole, M., & Yeoman, G. H. (1973) Prasinons A and B: potent insecticides from *Streptomyces prasinus. Appl. Microbiol.* **26**, No. 5, 699–704.

Box, S. J., Gilpin, M. L., Gwynn, M., Hanscomb, G., Spear, S. R., & Brown, A. G. (1983), MM 14201, a new epoxyquinone derivative with antibacterial activity produced by a species of *Streptomyces*. *J. Antibiot.* **36**, No. 12, 1631–1637.

Brown, A. G. (1981) New naturally occurring β-lactam antibiotics and related compounds. *J. Antimicrob. Chemother.* 7, 15–48.

Butterworth, D., Cole, M., Hanscomb, G., & Rolinson, G. N. (1979) Olivanic acids, a family of β-lactam antibiotics with β-lactamase inhibitory properties produced by *Streptomyces* sp. I. Detection, properties and fermentation studies. *J. Antibiot.* **32**, 287–294.

Butterworth, D., Hood, J. D., & Verrall, M. S. (1983) The new β-lactam antibiotics. *Advances in biotechnological processes* **1**, 251–292.

Hood, J. D., Box, S. J., & Verrall, M. S. (1979), Olivanic acids, a family of β-lactam antibiotics with β-lactamase inhibitory properties produced by *Streptomyces* sp. II. Isolation and characterisation of the olivanic acids MM 4550, MM 13902 and MM 17880 from *Streptomyces olivaceus*. *J. Antibiot.* **32**, No. 4, 295–304.

Hood, J. D. (1982) Inhibitors of antibiotic inactivating enzymes. In: Bu'lock, J. D., Nisbet, L. J., & Winstanley, D. J. (eds.) *Bioactive Microbial Products, Search and Discovery*. London, Academic Press, pp. 131–145.

Miller, T. W., Goegelman, R. T., Weston, R. G., Putter, I., & Wolf, F. J. (1972) Cephamycins, a new family of β-lactam antibiotics, II. Isolation and chemical characterisation. *Antimicrob. Ag. Chemother.* **2**, No. 3, 132–135.

Parker, W. L., Rathnum, M. L., Wells, J. S., Trejo, W. H., Principe, P. A., & Sykes R. B. (1982) SQ27,860, a simple carbapenem produced by species of *Serratia* and *Erwinia*. *J. Antibiot.* **35**, No. 6, 653–660.

Reading, C., & Cole, M. (1977) Clavulanic acid, a beta-lactamase inhibiting beta-lactam from *Streptomyces clavuligerus*. *Antimicrob. Ag. Chemother.* **11**, No. 5, 852–857.

Sykes, R. B., Cimarusti, C. M., Bonner, D. P., Bush, K., Floyd, D. M., Georgopapakou, N. H., Koster, W. H., Liu, W. C., Parker, W. L., Principe, P. A., Rathnum, M. L., Slusarchyk, W. A., Trejo, W. H., & Wells, J. S. (1981) Monocyclic β-lactam antibiotics produced by bacteria. *Nature* **291**, 489–491.

3

Integration of fermentation and recovery processes

A. R. Thomson
Biochemistry Group, Engineering Sciences Division,
Harwell, Oxfordshire OX11 0RA, UK

3.1 SUMMARY

The usual current practice in the biotechnology industry, is to develop the recovery and purification process required to fit in with an already optimised production process. The author argues that a much more integrated approach to development of the whole process is called for and is feasible, owing to recent developments in separations technology and genetic engineering. This view is illustrated by examples from published work.

3.2 INTRODUCTION

Microorganisms are able to produce a bewildering variety of compounds ranging from small molecules to high molecular weight polymers, and can act as hosts for the production of viral particles, some of which are used as vaccines. These products can be intracellular or extracellular. 'Fermentation' processes have well-recognised advantages for producing chemicals, but also have attendant disadvantages. Most of these are inherent in biological systems and include relatively low product concentrations, highly aqueous systems, complex mixtures, non-Newtonian process streams, and instability.

Until recently, the organism in which a desired product is produced naturally has been the determinant in a fermentation process. Thus attention has been focused almost entirely on maximising product yield. The recovery process has been developed and scaled-up subsequently, and further processing of reject streams has been regarded as a waste disposal problem. In essence, product recovery has been developed to fit in with an optimised fermentation.

The reasons why this pattern is common are not difficult to see, particularly for pharmaceutically active materials. Frequently it is necessary to determine first if a new compound is bioactive. If so, more is required to carry out initial toxicity and other tests, followed by more detailed tests to establish its full potency and range of actions as well as any indications of side effects or undesired contaminating activities. Not infrequently these latter tests require some regulatory approval which generally requires details of the method of preparation to be established and declared. It is only at this stage that approval for more widespread testing or use is obtainable. By this time, however, any significant change in the production method used is likely to require very substantial retesting at considerable expense. It is generally considered therefore not worthwhile to invest in significant process development until efficacy of the compound is proved. By this time changes in the production method may be quite costly.

This dichotomy between optimisation of production and of the recovery process though undesirable, is nevertheless common in spite of the fact that recovery and purification costs can amount to 80% of total production costs. Initially, in a product life cycle this may not be too serious a consideration, but as patent cover runs out and competition increases, reduction in production costs become increasingly important, even for relatively high value compounds.

Clearly there could be substantial benefits in integrating fermentation and downstream processing (Paul *et al.* 1981, Thomson 1984). For example, a number of workers have investigated the reactor/separator concept (Wandrey 1984); substantial advantages can be envisaged by avoiding the initial solid/liquid separation step (e.g. of cells from medium) (Atkinson & Sainter 1983, Wase *et al.* 1983), and the use of high resolution process steps (e.g. based on affinity) could minimise losses and speed up purification (Secher & Burke 1980).

The advent of recombinant DNA techniques has revolutionised the production of bioactive compounds – not least by the totally new possibilities for integrating fermentation with downstream processing. Since in principle, any protein can be produced in any chosen microorganism (and possibly in future in any plant or animal cell line), a new concept of integrated bioprocess can be envisaged. In this, the host organism will be selected to give maximal productivity with the minimum energy and chemical inputs, and to enable the simplest and most cost-effective recovery process to be used. Furthermore, the fermenter/bioreactor will be part of this overall optimisation, which will also enable the recovery of useful by-products to be maximised.

The technology required to bring this about already is or will be available, given adequate research priority. Indeed, research in both fermentation and separation technologies should be orientated to maximise the possibilities inherent in these technologies.

What then are these new possibilities, how can they be developed, and what are the likely benefits?

3.3 IN THE FERMENTER

3.3.1 Extending the range of producer organisms

The remarkable power of recombinant DNA technology is amply exemplified by the production of mammalian proteins in microbial cells. This power carries with it now a greatly enhanced freedom of choice in selecting the host organism. Relevant advantageous characteristics of a number of species are listed in Table 3.1. This could enable the genetic engineer to select a host most suitable for the product, taking into account recovery and purification. Thus *E. coli* might be the most suitable host for the production of basic proteins, since most of its own proteins have acidic isoelectric points (M. Bier, private communication). Thus, purification could be greatly simplified. This is illustrated in part by the example quoted below (Sassenfeld & Brewer 1984).

Table 3.1 – Potential host cells.

Cell type	Advantages	Disadvantages
Microorganisms		
E. coli	Genetics known, grows rapidly	Endotoxins, occasional human pathogen, no glycosylation
B. subtilis	Grows rapidly, used on industrial scale, secretes proteins, no endotoxins, non-pathogenic?	Secretes proteases, genetics little studied
Yeasts	Non-pathogenic generally, used as human food, flocculating strains, glycosylate, genetics well studied, grown on very large scale	Glycosylation different to mammalian system
Thermophiles	Fewer contamination problems, often more stable proteins	Genetics little studied, glycosylation not studied, pathogenicity not known
Cells from higher organisms		
Animal	Correct post-transcription modifications, glycosylates, antigenicity minimised, oxygen requirements low, proteins secreted	Genetics–vectors not well developed, possible carry-over of viruses/oncogenes, slow growth, expensive nutrients

Table 3.1 (cont.)

Plant	Safety *re* mammalian products, wide range of products, low oxygen requirements	Genetics complex, few vectors as yet, post-transcriptional modification system poorly known, glycosylation?

3.3.2 Increased product concentration

A major problem in the recovery and purification of 'natural' products is the fact that product concentrations tend to be low, <5%, often very much lower. Already by conventional mutation/selection techniques, 'runaway' production of proteins can be achieved by, for example, organisms such as *B. subtilis*, and *Asp. niger*. In some genetically engineered cells however, it has been possible to achieve levels of expression of foreign genes of up to ~20% of total cellular protein, and up to 40% when self-cloned (Goeddel *et al.* 1979, Miozzari 1981). Clearly this will simplify recovery and purification. It also enables production of rare proteins in much larger quantities than are available from natural sources, e.g. 30% of total protein (Masui *et al.* 1984).

3.3.3 Decoupling production from growth

Although production of many cell proteins parallel cell growth, it has been proved possible to decouple growth from product synthesis. This has been achieved by mutation and careful strain selection and by manipulating microbial cell physiology by modifying environmental conditions (e.g. limiting specific nutrients). More recently, recombinant DNA techniques have been used (Remaut *et al.* 1983). In one example, the gene coding for the desired protein product has been coupled to a gene which is expressed only at an elevated temperature. Thus growth of the cells (*E. coli*) is carried out under normal culture conditions, and production is then initiated and maintained by raising the temperature (Table 3.2). In this way, the product will be present in the fermenter for a shorter time (this could result in reduced proteolysis for example), and recovery could be simplified if it were possible to change to a new or simpler culture medium before switching on production.

3.3.4 Proteins as inclusion bodies

Although fortuitous and unexpected, there are clearly advantages in producing proteins in microorganisms as inclusion bodies, i.e. as precipitated granules. Many foreign genes (particularly those of eukaryotes) expressed at high levels in prokaryotes such as *E. coli* are produced as insoluble, heterogeneous, often disulphide-bridged granules, and often in a semi-denatured form. These are relatively dense and are generally insoluble. Although isolating pure, homogeneous and *active* proteins from such granules is proving a problem, the existence of the desired product in this form has major advantages – not least that separation from many of the soluble cell proteins is relatively straightforward (Harris 1983). When techniques for recovering fully active protein from inclusion bodies are perfected, it is likely that this phenomenon will provide an even more valuable and effective way of improving production efficiency, at least for proteins.

Table 3.2 – Thermoinduction of plasmid expression.†,‡

Plasmid	At 28°C		At 42°C	
	Plasmid copies per chromosome	% of total DNA	Plasmid copies per chromosome	% of total DNA
pPLc281ig8	38	5	42	5
pcI857	29	3	38	4
pCP40	52	7	599	45
pKN402	52	12	521	51
pCP39	64	8	833	53

† All bacterial strains used were *E. coli* K-12 derivatives.
‡ For full details see Remaut *et al.* (1983).

3.4 DOWNSTREAM

3.4.1 Separation of cells from medium

In virtually all processes for obtaining bioactive materials from biological sources, whether intra- or extracellular, an important stage is separation of the cells or cell fragments from medium or extracting fluid. Since this is almost invariably the initial step in product recovery, it is carried out on the largest scale. Consequently, substantial costs are incurred and even small increases in efficiency (or reduction in costs) could have a significant impact on the production costs (Atkinson & Sainter 1983). Furthermore, in view of the instability of biological feedstocks, minimal processing time is generally highly desirable, otherwise substantial cooling costs are incurred in stabilising the feedstock. In addition, for protein products, minimum exposure to proteases is essential – again requiring minimal process time.

The usual first step in solid/liquid separation in a laboratory and for high value compounds is centrifugation. This is very capital intensive, though highly effective. Its effectiveness is much less for small microorganisms such as *E. coli* because of their small size (~1–5 μm), and even less for sub-cellular organelles and particulates generated by disruption of microbial cells. Alternatives include filtration with or without added filter-aid, cross-flow filtration with membranes, flocculation and aggregation.

It is well known, however, that some strains of microorganism will flocculate naturally without any additives. Examples include some yeast strains (Miki *et al.* 1980) (Fig. 3.1). Research has shown that the flocculating characteristics are linked genetically and that in yeast the genes can be transferred from non-competent species by mating. More recently such transfers have been effected by recombinant DNA techniques. This opens out significant new possibilities for recovering cells minimising use of additives, such as flocculants, which for many medical products may cause regulatory difficulties. Additives always cost money and in general may need to be removed from the final product – resulting in additional costs.

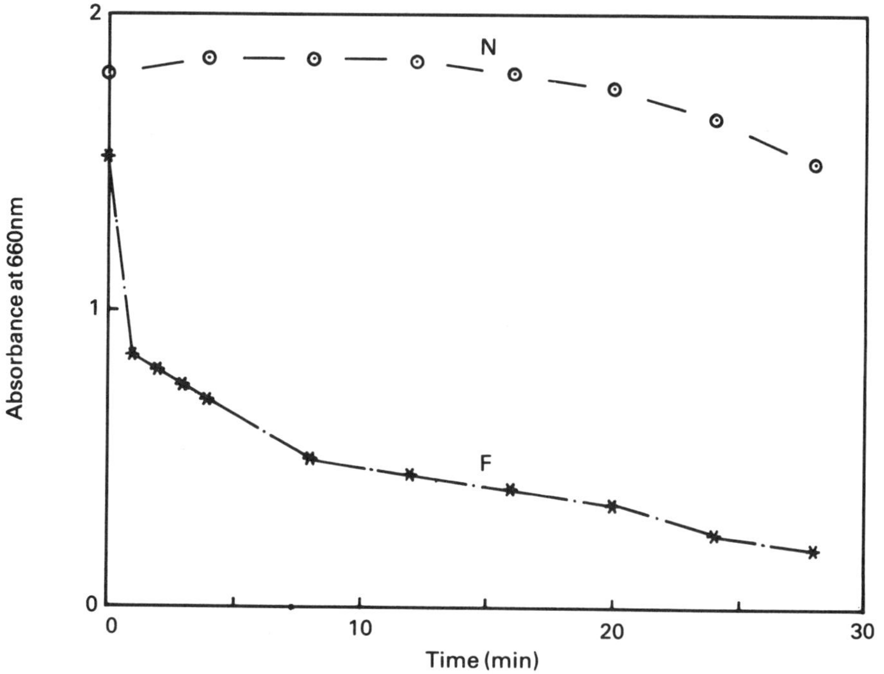

Fig. 3.1 – Sedimentation of yeast cells. Settling profiles for (F) flocculent (S656-1B) and non-flocculent (N) (S646-8D) strains of *Sacc. cerevisiae.*

The ICI Pruteen process illustrates another approach. Here, the cells are caused to aggregate by heat-treatment and adjustment of environmental conditions, and the bulk of the solid/liquid separation is carried out in a very large settling tank under gravity and under aseptic conditions, so that the clarified supernatant can be recirculated into the fermenter to maintain maximal use of medium consituents and to maximise cell concentration (Scott 1983). Heat treatment will be applicable to only a limited range of products because, for example, of denaturation in the case of proteins.

3.4.2 Direct broth extraction

Where the product is in solution in the culture broth (i.e. extracellular), it should be possible in principle to extract it without separating the cells from the medium. This is obviously a formidable task and considerable problems can be envisaged in carrying this out owing to the presence of cells and suspended solids, and further particulates may be produced by disruption of the cells in the fermenter or during subsequent processing. There may also be interfering solutes including ions present. Direct broth extraction is used however, in at least one major process, viz. streptomycin production, and many of the above problems have been encountered, and in this case at least partially overcome. Direct extraction is more attractive for streptomycin than many other antibiotics and small molecules because of its very high water solubility making solvent extraction unpromising. In the process, the whole broth is contacted with an ion-exchange resin and the antibiotic is recovered by subsequent elution (Bartels *et al.* 1958).

Recovery of proteins by direct broth extraction has hardly been looked into. Undoubtedly this is due in part to the fact that the feasible adsorbents such as ion-exchange celluloses, and polymerised dextran and agarose derivatives, have unpromising characteristics given the presence of cells and other particulates.

We at Harwell have supported research at Birmingham University on investigating the use of our fabricated porous, inorganic particulates for direct broth extraction† (Wase *et al.* 1983). *Bacillus cereus* in continuous culture produces penicillinase and cephalosporinase extracellularly. The research so far has shown that these enzymes can be adsorbed with high efficiency from unclarified whole broth, by a number of fabricated inorganic materials, including titania, kieselguhr, Fuller's earth and Bentonite (Table 3.3). Good recoveries of active enzyme can be obtained. This appoach is being explored further by us, with a view to integrating the recovery step directly with a continuous fermentation. Clearly there will be major problems in deploying the adsorbent effectively and efficiently in an operating process, but in principle solutions to these problems can be envisaged.

Table 3.3 – Direct broth extraction. Adsorption of penicillinase by fabricated inorganic adsorbents.‡

Adsorbent	Penicillinase adsorbed (%) Non-centrifuged broth	Penicillinase adsorbed (%) Centrifuged broth
Titania (TiO_2) (150–250μm)	92.9	97.4
Titania (~600μm)	93.0	96.0
Kieselguhr (~200μm)	93.6	95.4
Fullers earth (90–210μm)	88.2	86.9

† Now available from Sterling Organics, Cramlington, Northumberside NE23 7QG, UK.
‡ Uptake determined by batch adsorption, at pH of culture (pH 6–6.5), adsorbent equilibrated with phosphate buffer 0.1 M pH 7.0.

3.4.3 Cell disruption

A major step in the recovery of intracellular products is cell disruption. Although in the long term, it may be possible to develop specific methods for carrying this out in a controlled manner, it is more likely that physical and mechanical techniques for disrupting cells will be the most commonly used in the immediate future. It may however, be possible to develop techniques for disrupting the outer cells walls of microorganisms differentially, so allowing release of products concentrated in the periplasmic space without general release of intracellular components.

Recombinant DNA technology offers alternatives here also. Thus it has been possible to introduce genes which weaken the cell walls of bacteria, so that lysis occurs under much milder conditions than in the absence of those genes (Sinsky 1981). The possible advantage of this is obviously considerable, not only in increasing the efficiency and hence throughput of existing equipment such as cell homogenisers, but in the possibility of using low concentrations of chemicals or enzymes, alone or in combination with physical and mechanical techniques.

Although this approach has been stimulated by the advent of rDNA techniques, it is clear that it could be developed more widely and perhaps more simply by conventional mutation and selection procedures. Ease of cell disruption could and should therefore be included among the criteria for selecting a particular organism as a production strain.

3.4.4 Product secretion

Until recently, the bulk of industrial enzymes have been extracellular. Recovery has involved removal of cells from the medium and then concentration of the clarified broth, with some purification. Intracellular proteins have generally been produced on a much lower scale, with the notable exceptions of invertase and glucosidase (Godfrey & Reichelt 1983). Production of the latter two enzymes has involved cell disruption, separation of debris and particulates from the extract and further purification of the desired product from the many other proteins in the extract. This is the general pattern for intracellular enzymes. Clearly there are major advantages in the desired product being extracellular. rDNA techniques are now being used increasingly to produce proteins of medium to high value, and on an increasing scale. As indicated previously, there are major advantages in retaining products within the cell, particularly if they are produced in the form of inclusion bodies, since substantial concentration is achieved. However, yields of active protein obtained from such inclusions are often low since processing generally involves the use of denaturing and reducing conditions to disaggregate the heterogeneous body, followed by purification in the presence of denaturants, and finally renaturation. The latter may be even more of a problem if the protein product is composed of different sub-units, perhaps linked by disulphide bridges.

In view of these considerations therefore, secretion of product into the medium could well in the end be more advantageous. Two ways have been investigated recently for doing this:

(i) by alteration of the salt balance of the medium (Kato *et al.* 1983). Thus it has been found that increasing K^+ concentration in the medium used for growth of *E. coli* can lead to substantial secretion of proteins into the culture fluid (Table 3.4). This involves no genetic manipulation. It is not known how widespread this phenomenon is, or if its use can be sustained for a significant period during a fermentation, or if it acts as one-off trigger mechanism releasing a burst of intracellular product. Nor is it known whether it is specific for certain products.

(ii) A much more specific technique has been developed using genetic manipulation. Much research has been devoted to understanding protein transport in microbial cells, as it has been known for some time that secretion of proteins into culture media is highly specific. A favoured mechanism proposed for secretion is as follows (Talmadge *et al.* 1980). The gene for the naturally secreted protein includes the code for a 'header' sequence which when the protein is expressed in the cell binds to components in the cell membrane and mediates transport through it. Before emergence into the periplasmic space of the exterior of the cell through the cell wall, this header peptide is cleaved yielding the secreted product protein in its native form. The sequence(s) of header peptides have been determined in some microorganisms and encoded in inserted genes. It has indeed been found in this way that export of foreign proteins can be induced. An example of this is interferon produced in *E. coli.* The header sequence is attached to the encoded gene for interferon and interferon without header peptide is found in the medium Table 3.5 (Palva *et al.* 1983). No interferon is extracted if the header sequence is absent from the introduced gene, even though this gene is expressed to a significant level in the cell.

Table 3.4 – Effect of metal ions on secretion of penicillinase by *E. coli.*†

Salts added	Concentration	Extracellular	Periplasmic	Cellular‡	Total
None		1.2	0.1	7.8	9.1
NaCl	0.08 M	11.9		4.2	16.1
	0.16 M (1%)	20.5	0.2	2.5	23.2
	0.32 M	14.1		3.8	17.9
	0.80 M	2.5		3.5	6.0
Na_2SO_4	0.08 M	16.3		5.3	21.6
KCl	0.16 M	16.2		4.8	21.0
LiCl	0.16 M	16.8		3.0	19.8
CsCl	0.16 M	0.1		0.1	

† *E. coli* HB101 containing plasmid pEAP2 with alkalophilic *Bacillus* No. 170 penicillinase gene. From Kato *et al.* (1983).

‡ Intracellular activities. Media used: the basic + 0.1% glycerol salts. Values = units of penicillinase activity/ml culture.

Table 3.5 – Secretion of interferon synthesised in transformed *B. subtilis*.†

Transforming plasmid	Interferon activity (IU/ml original culture)	
	Secreted	Cell associated
pKTH68	$1\text{–}2 \times 10^5$	6×10^3
pKTH93	1×10^5	6×10^3

† *B. subtilis* IH 6140 was used and was transformed with either of the above plasmids (for more details see Palva *et al.* (1983)).

3.4.5 Reactor/separator concept

Direct broth extraction as described above (Wase *et al.* 1983) and the ICI Pruteen plant (Scott 1983), in both of which there is separation of product closely integrated with fermentation, are examples of the reactor/separator concept. Indeed in the case of the Pruteen plant the product (biomass in the form of microbial cells) is separated after conditioning the whole broth process stream by heating and adjustment of the medium constituents, in an aseptic settling tank within the aseptic containment enclosure of the fermenter. Vacuum fermentation (being developed for ethanol production), is another example of closely integrating the downstream processing with a fermenter, although the driving force in this case is largely to reduce distillation costs (Finn & Ramalingam 1977, Cysewski & Wilke 1977). A major object of process design in the Pruteen plant is recycling of medium. In other configurations recycle of cells is used to maximise cell density, and hence process intensity (Cysewski & Wilke 1977). Clearly this also requires a separation step closely integrated with the fermenter.

3.4.6 Immobilised biocatalysts

Much research has been devoted to immobilising enzymes, and major industrial processes depend on these. A notable example is high fructose syrup produced on 1M ton/annum scale in the USA. Immobilisation technology is being applied increasingly to whole microbial and also plant and animal cells. Although re-use of the catalyst is a major advantage of immobilisation, there can be a number of beneficial effects on down stream processing (Atkinson & Sainter 1983, Rosevear & Lambe 1983). For example, since the catalyst (e.g. cells) is immobilised and precursors and products diffused into and out of the system, no solid/liquid separation step is required in isolating products. Furthermore, there is minimal contamination of the product stream with particles such as cell debris not uncommon in fermenter broths. These are derived from cell lysis due to death or damage from the mechanical stirrers usually required for adequate oxygen transfer. Consequently, the product stream is analogous to very efficiently clarified broth. It is known that a major constraint in processing broths by ion-exchange is the presence of fine particulates and colloidal materials. It is probable that these can be minimised in an immobilised bio-catalyst system, so that an extended operating life can be anticipated (A. Rosevear & S. Roe, unpublished observations).

It may well be, therefore, that these advantages will act as a stimulus when considered in conjunction with others such as catalyst re-use, compactness of plant and possibly substantial reduction in capital costs, in the further development and more extensive use of immobilised bioreactor systems.

3.4.7 Gene modification

Recombinant DNA technology of gene splicing and gene synthesis has been utilised over the past year in an uniquely new way to facilitate recovery of polypeptides and proteins (Sassenfeld & Brewer 1984). The approach used is exemplified in the production and isolation of human β-urogastrone. This protein differs from γ-urogastrone in having a C-terminal arginine. It also has a neutral/acidic pI. A DNA sequence coding for five additional arginines at the carboxyl terminus was inserted into the β-urogastrone gene. This gene fusion resulted in the production of β-urogastrone in *E. coli* as a modified, highly basic protein. The harvested cells were disrupted by sonication, and the extract was fractionated by ion-exchange chromatography on a sulphated cation exchanger. The eluted polyarginyl-urogastrone was then treated with the exopeptidase carboxypeptidase A which removed basic residues such as arginine from the C-terminus of the polypeptide chain. This treatment alters substantially and specifically the charge on the polypeptide. Consequently, when re-chromatographed on the same exchanger, the digested and hence much less basic product eluted at a totally different position in the gradient and was thus resolved from the previously cochromatographing contaminants (Fig. 3.2). This principle appears to be simple,

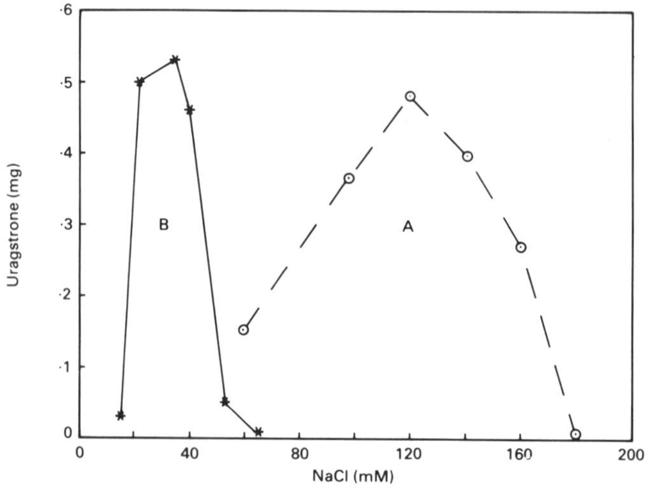

Fig. 3.2 – Purification of urogastrone. Ion-exchanger used: SP-Sephadex. Peak A, polyarginylurogastrone; Peak B, peak A digested with carboxypeptidase B and rechromatographed on SP-Sephadex under the same conditions.

effective and suitable for scale-up, as well as being applicable in principle to any recombinant protein. It illustrates beautifully the potential of recombinant DNA technology, both for simplifying downstream processing and for process integration.

3.5 CONCLUSIONS

There seems little doubt that substantial benefits can be gained from close integration of production and recovery processes, in spite of the difficulties inherent in new biological products and in the organisation of the R&D required. The broadening scope of conventional biotechnology and in particular the development of recombinant DNA techniques, open up new possibilities for doing this effectively. In this chapter, I have discussed a number of ways in which this can be done. It is certain that additional and even more effective approaches to integrating recovery with the production process will be developed as the range of products expand and competition increases in the market place.

ACKNOWLEDGEMENTS

The author wishes to acknowledge with gratitude the support of the Chemicals, Minerals and Vehicles Requirements Board of the Department of Trade and Industry, and the Biotechnology Unit of the Laboratory of the Government Chemist. The views expressed however are those of the author.

REFERENCES

Atkinson, B., & Sainter, P. (1983) *Technological Forecasting for Downstream Processing in Biotechnology, Phase 2; Process and Unit Operation Needs,* FAST Series 7. Report EUR 8549EN.

Bartels, C. R., Kleiman, G., Korzun, J. N., & Irish, D. B. (1958) *Chem. Eng. Progr.* **54**, 49.

Cysewski, G. R., & Wilke, C. R. (1977) *Biotechnol. Bioeng.* **19**, 1125.

Finn, R. K., & Ramalingam, A. (1977) *Biotechnol. Bioeng.* **19**, 583.

Godfrey, T., & Reichelt, J. (1983) *Industrial Enzymology – The Application of Enzymes in Industry.* London, The Nature Press, Macmillan Publishers.

Goeddel, D. V., Kleid, D. G., Bolivar, F., Heynecker, H. C., Yansura, D. G., Crea, R., Hirose, T., Kraszeuski, A., Itakura, K., & Riggs, A. D. (1979) *Proc. Natl. Acad. Sci., USA* **76**, 106.

Harris, T. J. R. (1983), In: *Genetic Engineering.* Vol. 4. New York, Academic Press, p. 127.

Kato, C., Kudo, T., Watanabe, K., & Hirikoshi, K. *(1983) Eur. J. Appl. Microbiol. Biotech.* **18**, 339.

Masui, Y., Mizuno, T., & Inouye, M. (1984) *Biotechnology* January, 81.

Miki, B. L. A., Poon, N. H., James, A. P., & Seligy, V. L. (1980) In: *Current Development in Yeast Research*. Advances in Biotechnology, Proc. International Yeast Symp., London, Ontario, p. 193.

Miozzari, G. F. (1981) In: Gueriguian, J. L. (ed.), *Insulin, Growth Hormone and Recombinant DNA Technology*. New York, John L. Raven, p. 13.

Palva, I., Lehtovaara, P., Kaarianinen, L., Sibakov, M., Cantell, K., Schein, C. H., Kashiwagi, K., & Weissmann, C. (1983) *Gene* **22**, 229.

Paul, E. L., Kaufman, A., & Sklarz, W. A. (1981) *Ann N.Y., Acad. Sci.* **369**, 181.

Remaut, E., Tsao, H., & Fiers, W. (1983) *Gene* **22**, 103.

Rosevear, A., & Lambe, C. A. (1983) In: Wiseman, A. (ed.), *Topics in Enzyme and Fermentation Biotechnology*, Vol. 7. Chichester: Ellis Horwood, p. 1.

Sassenfeld, H., & Brewer, S. J. (1984) *Biotechnology*, January, 76.

Scott, R. (1983) *Biotech '83*. Northwood, UK, Online Publications, p. 235.

Secher, D. S., & Burke, D. C. (1980) *Nature (London)* **285**, 446.

Sinsky, A. (1981) In: *Advances in Fermentation Recovery Process Technology*. Engineering Foundation Conferences, Banff, Canada.

Talmadge, K., Stahl, S., & Gilbert, W. (1980), *Proc. Natl. Acad. Sci.* **77**, 3369.

Thomson, A. R. (1984) In: *Recovery of Fermentation Products*. Engineering Foundation Conferences, Sea Island, Georgia, USA, January 1984.

Wandrey, C. (1984) In: *Biotech '84 Europe*. Northwood, UK, Online Publications, p. 391.

Wase, D. A. J., Green, S., & Thomson, A. R. (1983) Proc. Conf. on Advances in Fermentation, Rickmansworth, Herts, Wheatland Journals Ltd, Supplement to *Proc. Biochemistry* p. 54.

4

Protein recovery from unclarified cell homogenates

Per Hedman, Jan-Gunnar Gustafsson and Ann-Kristin Frej
Pharmacia AB, S-751 82 Uppsala, Sweden

4.1 INTRODUCTION

When an intracellular protein is recovered from a fermentation broth the total volume is reduced during the cell harvesting step. The necessary cell disruption or lysis that follows will mix all cell constituents thoroughly. Proteolytic enzymes will degrade soluble products more easily than insoluble ones, and the commonly used protease inhibitors do not appear to be effective when added to crude, unclarified cell homogenates. Using a sulphhydryl group-labelled, denatured and reduced albumin substrate (Hedman & Gustafsson, in preparation) we measured remaining activities and found considerable proteolytic activity in the presence of the inhibitors used (Table 4.1). A limited screening of chromatography gel media brought our attention to hydrophobic gels e.g. Phenyl Sepharose® CL-4B and Octyl Sepharose CL-4B which bind much of the proteolytic activity in *E. coli* and bakers yeast (Hedman & Gustafsson 1983) (Fig. 4.1). The time delay between disintegration of cells until the proteases are bound and binding of proteases may however, result in considerable degradation of the product. This is due to the difficult clarification of the cell homogenate on a large scale.

An improvement in product stability is possible if the product is rapidly adsorbed to a solid. Exclusion or steric hindrance effect may keep the degrading enzymes away from susceptible bonds. In a well-mixed reactor, product binding may be a rapid procedure.

The recovery problem is now converted to a solid–solid separation problem. Solids of different kinds may be separated for example by their different behaviour in magnetic fields (Whitesides *et al.* 1983), different rates of sedimentation or by differences in partition coefficients in an aqueous two-phase extraction.

Albertsson (1971) has described cell and organelle separations based on partitioning in aqueous two-phase systems. A synthetic adsorbent can be designed to have a partition coefficient very different from that of the cell fragments. It is thus possible to separate the adsorbent from cell debris in a wide range of phase system compositions. Using a suitable protein binding ligand, and a two-phase system it is possible to remove a protein from both the cell debris and some soluble impurities in a two-phase system. The phase separation time depends on the system used, generally being shorter with polyethyleneglycol–phosphate systems than with polyethyleneglycol–dextran systems. Separation by gravity is accomplished within 10–30 min. If a low speed centrifugation (2000 g) is used, the phase separation is complete within a minute.

Table 4.1 – Effect of inhibitors on proteases in a yeast homogenate.

Inhibitor	Protease (cpm)	Activity (% of control)
Phenylmethylsulphonylfluoride 10 mM (suspension)	13 400	75
Benzamidine 10 mM	14 500	81
N-Ethylmaleimide 10 mM	15 400	86
Ethylenediaminetetraacetate 10 mM	14 500	81
Potato Tuber Protease Inhibitor A 13 mg ml^{-1}	10 900	61
Potato Tuber Protease Inhibitor B 14 mg ml^{-1}	13 400	75

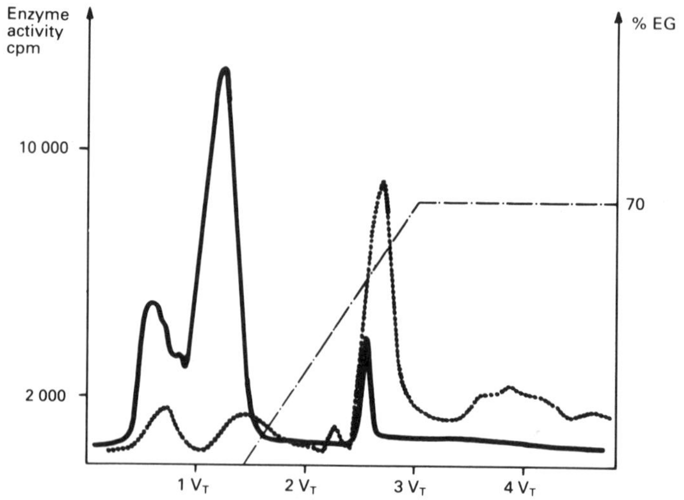

Fig. 4.1 – *E. coli*, 4°C, 0.5 M NaCl in 50 mM sodium phosphate. ——— A_{280}, ········ Enzyme activity, –·–·– Ethyleneglycol gradient. Octyl Sepharose CL-4B.

4.2 RESULTS AND DISCUSSION

To obtain suitable partitioning a number of substituents may be used (Hedman & Gustafsson 1984). We found that Phenyl Sepharose CL-4B has a suitable partition coefficient. In a test of our ideas we mixed this hydrophobic chromatography gel with an *E. coli* homogenate obtained from a glass bead mill.

When the *E. coli* proteins were in contact with this adsorbent in the presence of potassium phosphate, polyethyleneglycol (PEG) was added. The resulting phase system made it possible to drain off the cell fragments in the phosphate rich bottom phase. The top phase containing Phenyl Sepharose was poured into a column where the gel was washed in a high salt buffer. Two protein peaks were eluted, the first in a low salt buffer, the second with an ethyleneglycol gradient (Fig. 4.2). Polyacrylamide gel electrophoresis revealed that a number of proteins were obtained in each peak, some of them concentrated.

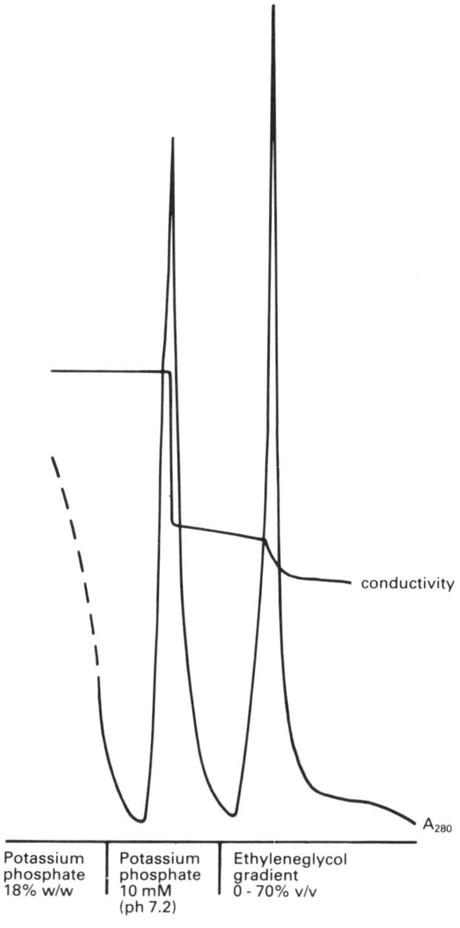

Fig. 4.2.

The successful use of the technique described above depends of course on the economics. The adsorbent recovery was thought to be one major factor affecting product yield as well as adsorbent cost. It is necessary to re-use the adsorbent to keep costs down. The economy of this process will also depend on the product concentration in the cell homogenate, that is, the expression level.

4.3 PROCESS ECONOMY

With Cibacron Blue substituted beads, it was possible to estimate the conditions necessary for adsorbent recovery. Based on our experiments, we consider losses of 0.5–2% adsorbent per cycle realistic. The number of cycles depends on the regeneration protocol. Forty cycles seems to be a realistic figure, given that no binding capacity loss could be measured after 10 cycles. The phase system components will also add to the total cost, Dextran being the most expensive and phosphate the least expensive by weight.

Figs. 4.3 and 4.4 show the result of cost calculations based on the assumptions above (0.5% adsorbent loss in Fig. 4.3, 2% in Fig. 4.4). The protein binding capacity which is necessary at different expression levels is within the range of available chromatography gel media. In Fig. 4.3 it is assumed that the product and 5% of the soluble contaminants bind to the gel. In Fig. 4.4 the binding is supposed to be less specific (one-third of the soluble contaminants and the product bind to the gel).

These calculations are merely intended to identify those factors which are most important for the economy. Obviously, low-cost alternatives to dextran, re-usable beads, high binding capacity and selectivity will all be important. For example, with a high capacity, less gel is needed. The inexpensive phosphate–PEG systems cannot be used with triazine dye ligands or ion exchanger beads. Thus it is desirable to find suitable low-cost polymers.

4.4 PRODUCT RECOVERY

In order to estimate protein recoveries we used Cibacron Blue ligand beads. Nucleotide cofactor-dependent enzymes are known to bind to Cibacron Blue at low salt concentrations. Glucose 6-phosphate dehydrogenase (G6PDH), a nucleotide cofactor-dependent enzyme, was eluted with a salt gradient and the effluent was assayed for enzymatic activity (Fig. 4.5). Obviously a certain degree of purification can be achieved during the elution. In a separate experiment the Cibacron Blue ligand beads were recovered from the PEG-rich phase in a PEG-dextran phase system after adsorption of proteins from a yeast homogenate. The recoveries of alcohol dehydrogenase, hexokinase and glucose 6-phosphate dehydrogenase after elution with 1.5 M potassium chloride were closely equal to those reported for Blue Sepharose CL 6B chromatography (Pharmacia).

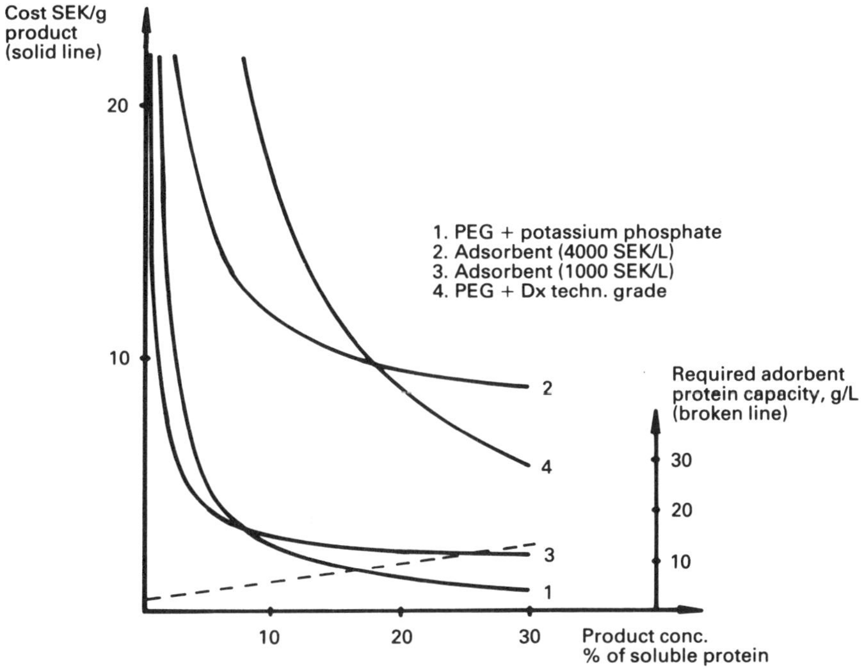

Fig. 4.3 – Cost analysis. Specific protein binding. Phase system contains 30% cell homogenate and 20% adsorbent. The homogenate contains 30 g/l soluble protein.

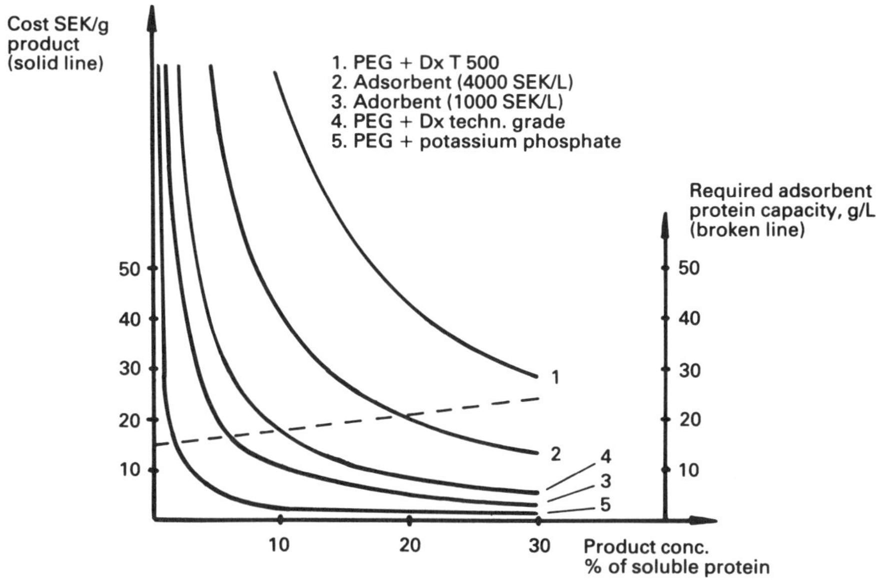

Fig. 4.4 – Cost analysis. Unspecific protein binding. Phase system contains 30% cell homogenate and 20% adsorbent. The homogenate contains 30 g/l soluble protein.

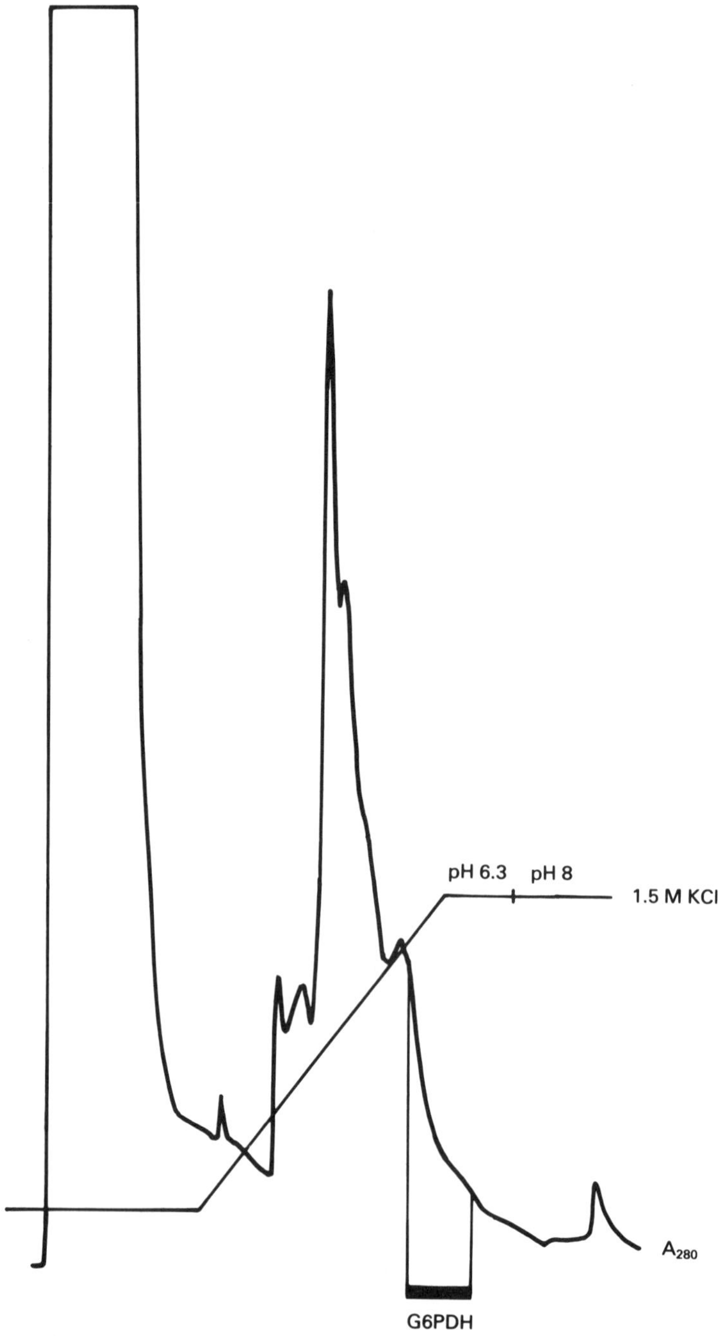

Fig. 4.5 – DNAse treated yeast homogenate components separated on a Cibacron Blue–PEG–Sepharose column.

We concluded that the phase partitioning *per se* did not affect the enzyme recovery. Still we were concerned with our moderate recoveries as compared to those reported by Kroner *et al.* (1982). It may be that our cell homogenates contained higher than optimal salt concentrations or that our system contained a higher level of competing impurities. We noticed that Pedersen *et al.* (1978) in their investigation of the relative abundance of *E. coli* proteins, found catabolic enzymes to be present at only low to moderate levels.

Utilizing a highly resolving ion exchanger FPLC system we found that the material eluted from the Cibacron Blue adsorbent contained a number of proteins (Fig. 4.6). We observed that ADH and G6PDH only constitute a small fraction of the protein bound to the adsorbent. This may explain our modest recoveries with Cibacron Blue–PEG–Sepharose.

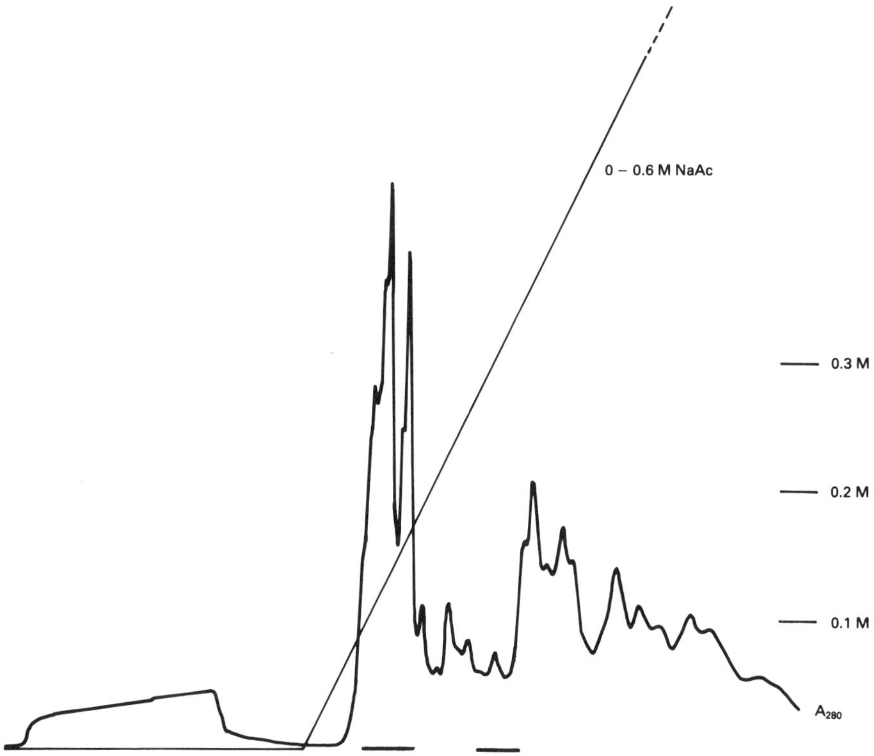

Fig. 4.6 – Separation of yeast proteins eluted with 1.5 M KCl from Cibacron Blue-PEG-Sepharose after partitioning in a 5% PEG 4000, 5% Dx T 500 phase system. Column: Mono Q (1 ml). Sample: Sephadex G25 desalted (10 ml) pH 7. Separation time: 35 min.

This means that a considerable fraction of the gel's binding sites are occupied by proteins other than the enzymes mentioned above. At high protein loading, the adsorption is best described by Langmuir isotherms (data not shown). It follows that, with few unoccupied binding sites left, the recovery of the selected enzymes is limited, unless a large excess of adsorbent is added. Consequently, it is important to use an adsorbent of appropriate specificity or a substantial excess of adsorbent will be needed. This is true for any batch adsorption process where contaminants may bind to the adsorbent.

Adsorption of the product protein immediately after homogenisation should improve the yield by reduction of the proteolytic breakdown of product. We found evidence in support of this idea when β-galactosidase was purified from an *E.·coli* homogenate in a two-step procedure (Frej *et al.* 1984). In the first step, the enzyme was bound to Phenyl Sepharose CL-4B. The beads were removed from the cell debris in a PEG–phosphate phase system. The β-galactosidase fraction eluted at low ionic strength (Fig. 4.7) was further purified on DEAE Sepharose Fast Flow (Fig. 4.8). The β-galactosidase content was monitored using FPLC during the process (Figs. 4.9–4.11). When the final yield (6.92 A/min from 592 g wet cells) was compared to the enzyme activity in a homogenate supernatant sample (kept at 4°C until analysis) we observed 124% 'recovery'. This means that during the processing time the enzyme activity in the starting material sample lost 20% of its β-galactosidase activity relative to the material which was adsorbed and further purified.

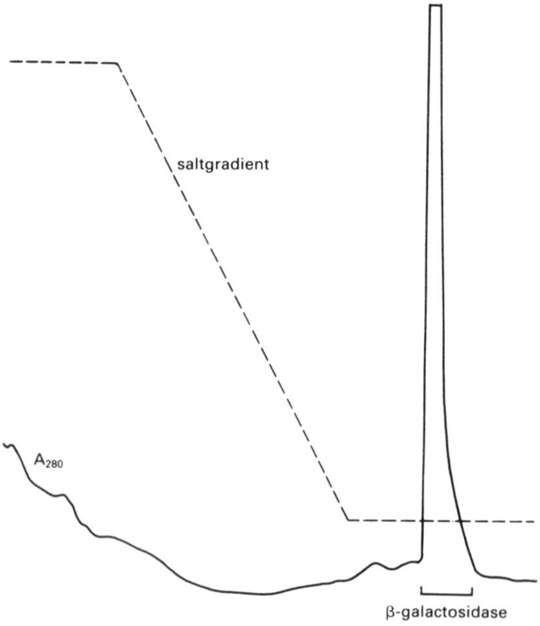

Fig. 4.7 – Elution of Phenyl Sepharose CL-4B. Gradient from 14% (w/w) potassium phosphate pH 7 to 3 mM potassium phosphate, 1 mM magnesium chloride 5 mM 2-mercaptoethanol, pH 7. Gradient volume 41. Flow rate 23 cm/h.

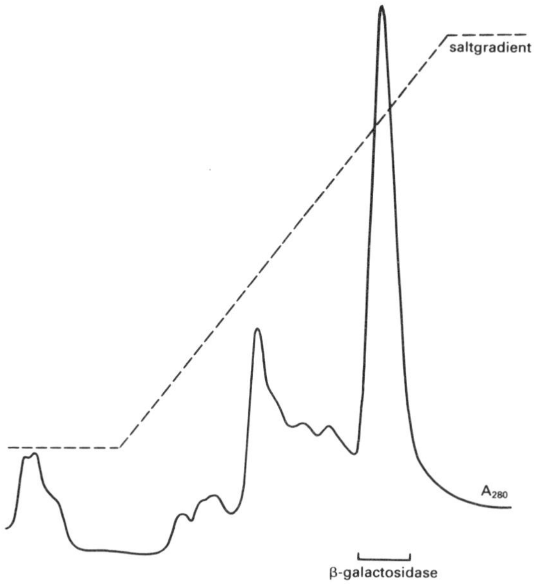

Fig. 4.8 – Chromatography on DEAE–Sepharose Fast Flow. Gradient 0–0.6 M sodium chloride in 50 mM Bis-Tris, 1 mM magnesium chloride and 5 mM 2-mercaptoethanol, pH 5.8. Gradient volume 8 l. Flow rate 23 cm/h.

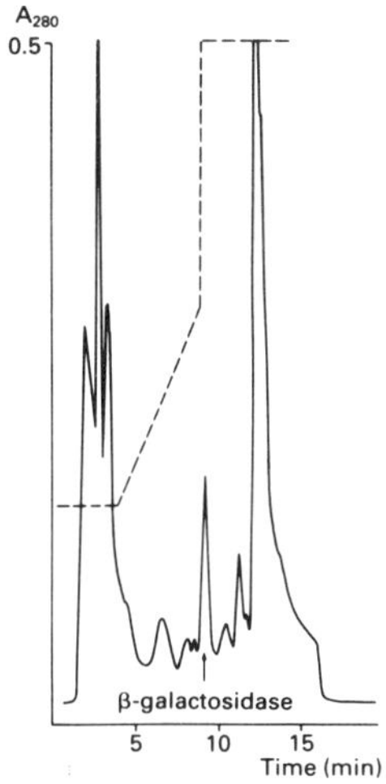

Fig. 4.9

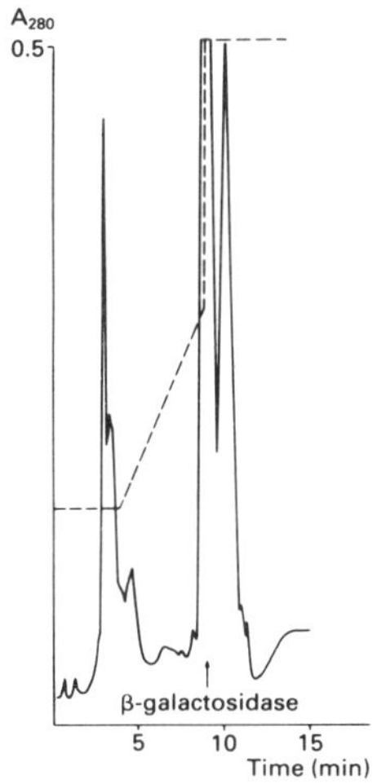

Fig. 4.10

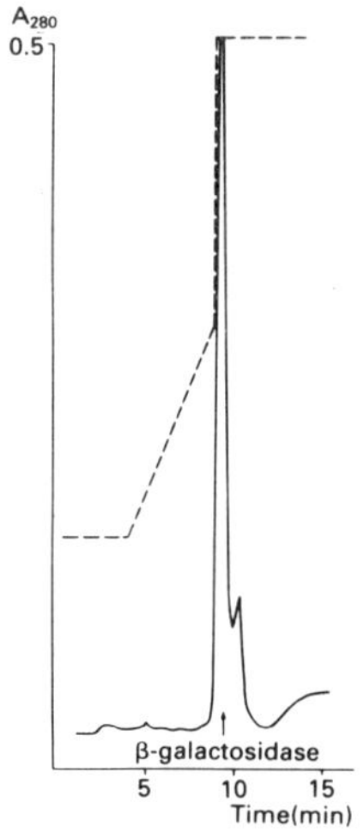

Fig. 4.11

Fig. 4.9 – Chromatography of cell homogenate supernatant. on a Mono QHR 5/5 prepacked column.

Fig. 4.10 – Chromatography of the combined β-galactosidase containing fractions from the Phenyl Sepharose CL-4B on a Mono QHR 5/5 prepacked column.

Fig. 4.11 – Chromatography of the β-galactosidase containing fractions from the DEAE–Sepharose Fast Flow on a Mono QHR 5/5 prepacked column.

4.5 CONCLUSIONS

Rapid recovery procedures based on a combination of adsorption and extraction may result in increased recoveries. The economic feasibility depends, of course, on the availability of both inexpensive phase-forming chemicals and suitable adsorbents. Since the adsorbents are in fact used for chromatographic purification in the elution step, their cost contribution should be acceptable where chromatographic methods are used.

REFERENCES

Albertsson, P. Å. (1971) *Partition of Cell Particles and Macromolecules*, 2nd edn. New York, John Wiley and Sons.

Frej, A.-K., Gustafsson, J.-G., & Hedman, P. (1984) Poster Presentation No. 149, Third European Congress on Biotechnology, Munich.

Hedman, P., & Gustafsson, J.-G. (1983) Reduction of Proteolytic Breakdown in Microbial Homogenates. Presented at the joint IABS/WHO meeting, Geneva, 29 November to 1 December. Submitted to *Journal of Biological Standardisation.*

Hedman, P., & Gustafsson, J.-G. (1984) *Analytical Biochemistry* **138**, 411–415.

Hedman, P., & Gustafsson, J.-G. Manuscript in preparation.

Kroner, K. H., Cordes, A., Schelper, A., Morr, M., Bückmann, A. F., & Kula, M.-R. (1982) In: Gribnau, T. C. J., Visser, J., & Nivard, R. J. F. (eds.) *Affinity Chromatography and Related Techniques.* Amsterdam, Elsevier.

Pedersen, S., Bloch, P. L., Reeh, S., & Neidhardt, F. C. (1978) *Cell* **14**, 179–190.

Pharmacia. *Blue Sepharose CL-6B for Affinity Chromatography.* Booklet available free of charge.

Whitesides, G. M., Kazlanskas, R. J., & Josephson, L. (1983) *Trends in Biotechnology*, **1**, No. 5, 144–148.

5

Applications of ultrafiltration in the recovery of microbial products

Anders Törnqvist
Alfa-Laval AB, Food and Dairy Engineering,
Box 64, S-22100, Lund, Sweden

5.1 INTRODUCTION

Separation by ultrafiltration has been a recognised application of membrane technology for many years. However, it was about ten years ago, when the first synthetic membranes became available, that ultrafiltration for industrial use became significant. Today, as a result of the development of these resistant and rugged membranes, ultrafiltration has become a generally accepted and prominent separation technique in many branches of industry.

In a number of application areas, ultrafiltration has become an integral and indispensable part of the process. In many instances, ultrafiltration yields better product qualities, especially in the manufacture of products that are easily degraded, such as enzymes, antibiotics and other biochemical products.

Rising oil prices, have made ultrafiltration an increasingly profitable alternative to other more energy-intensive methods, such as evaporation and spray-drying.

5.2 PRINCIPLE

Ultrafiltration is a purely physical separation process. In principle, it is a straining process, where the filter medium has pores so fine that substances with a molecular weight of a few thousands are retained, while smaller molecules and water can pass through the membrane.

In conventional filtration, all the fluid is forced through the filter medium, and a filter cake is formed ('dead end' filtration). When attempting to filter the substances involved in ultrafiltration, sub-micrometre-thick deposits would form, presenting an impenetrable barrier to filtration. The concentrated phase must, therefore, be continuously removed to keep the pore passages open. This is done by allowing the fluid to flow rapidly across the surface (cross flow) while a small amount passes through the membrane. This prevents concentrations that are too high from forming at the surface of the membrane and flow resistance through the membrane is reduced considerably.

5.2.1 Membrane

The ultrafiltration membrane consists of an active layer which is about $1/10\,\mu m$ thick, with pores in the range of 1.0 to 10 nm. This active layer, is supported by a thicker porous layer with very small resistance to flow.

In the membranes used in most industrial-scale equipment today, both layers are made of the same material, the differing porosities in these layers being made by special procedures during the manufacturing process. A new type of membrane, fast gaining acceptance, is the thin film composite membrane (TFC). In TFC membranes the active layer and thicker porous support layer are made from different materials.

The membrane itself sometimes needs a separate support to allow it to withstand the pressures involved in ultrafiltration, normally from 1 to 10 bar. This can consist of a separate structure in the form of a paper backing, or a porous material such as epoxy-reinforced glass fibre, which is the case with flat membranes and large-bore tubular membranes. The support structure can also be integrated with the membrane itself as with small-bore hollow fibre membranes.

Because the active layer of the membrane is so thin, the necessary pressure drop across the ultrafiltration membrane is moderate. In general, a pressure difference across the membrane of 1 to 3 bar is sufficient to drive the filtration process.

5.2.2 Membrane material

The first ultrafiltration membranes, were produced from cellulose acetate. As a result of the narrow pH tolerance and low temperature resistance of cellulose acetate, it soon became apparent that a better membrane material had to be developed.

The second-generation membranes, which have been on the market for nearly ten years, are made from such synthetic polymers as polysulphone, polyamide and other copolymers. These membranes have higher capacities and much wider pH tolerances. In general, they also tolerate higher temperatures and have greater resistance to solvents and other chemicals.

Even if the product to be separated does not attack the membrane material, and the process temperature is moderate, it is important, when selecting a membrane, to consider whether it tolerates the necessary cleaning solutions and temperatures, required for cleaning the membranes. In pharmaceutical and biochemical applications, the membrane must also tolerate disinfectants such as hypochlorite.

Most membrane manufacturers today, offer various membrane materials for various applications.

5.2.3 Membrane configurations

There are a variety of membrane configurations commercially available today.

(1) Plate and frame systems, having flat membranes stacked in a plate configuration, including necessary supports.
(2) Tubular systems, where the membranes are formed tubes, normally 13 to 25 mm in diameter. Supporting metallic or plastic tubes are incorporated in the system.
(3) Spiral systems in which the membranes are formed like Swiss rolls. The fluid is fed into one end of the roll and the concentrate comes out at the other end, while the permeate is discharged through a central tube.
(4) The hollow fibre system, consisting of small tubes with an inside diameter of 0.5 to 2 mm, is unique in the sense that they are self-supporting. The inner wall, which is the active surface, is integrated with the support structure. A large number of fibres are potted into both ends of a shell fitted with permeate ports.

Hollow fibre membranes have the unique property of withstanding pressure from either direction. Therefore they can be backflushed, which gives an additional cleaning option. For some applications this backflush facility is used on a regular basis but more normally its value is as a method for recovering membrane performance.

There is a small number of other configurations, which do not fit into the above classification. The best known is the leaf type, which is similar to a tubular type, but with a rectangular cross-section, with the membranes on leaves, or fins, projecting from one of the surfaces. In a later modification, this has become more like small rectangular plate and frame bundles, without gaskets between the membrane pairs, with the bundles placed into the rectangular cross-section flow path.

5.2.4 Scaling up and dimensioning

A UF plant is most often dimensioned for a new product on the basis of a pilot test. It is very important that the pressure, flow and concentration parameters of the full-scale system are reproduced during the pilot test.

The time factor is often overlooked. In general, flux declines with time, even if the concentration is constant. This effect is primarily attributable to deposits on the membrane surface.

5.3 METHODS OF OPERATION

5.3.1 Batch operation

In batch ultrafiltration (Fig. 5.1), the liquid is pumped from the batch tank, to the ultrafiltration unit. All or part of the concentrate is returned to the batch tank, the balance going back to the suction side of the main recirculation pump. The level in the batch tank falls at a rate determined by the amount of permeate produced, while the concentration level increases. Filtration continues until the desired concentration or volume is reached in the batch tank. The process is then interrupted, and the concentrate is drained off.

When filtering very dilute liquids, it is often preferable to first let the filtration take place, while continuously topping up the batch tank, until the concentration reaches the level, at which the flux begins to decline. The topping up of the batch tank is then stopped, and operation continues according to normal simple batch operation.

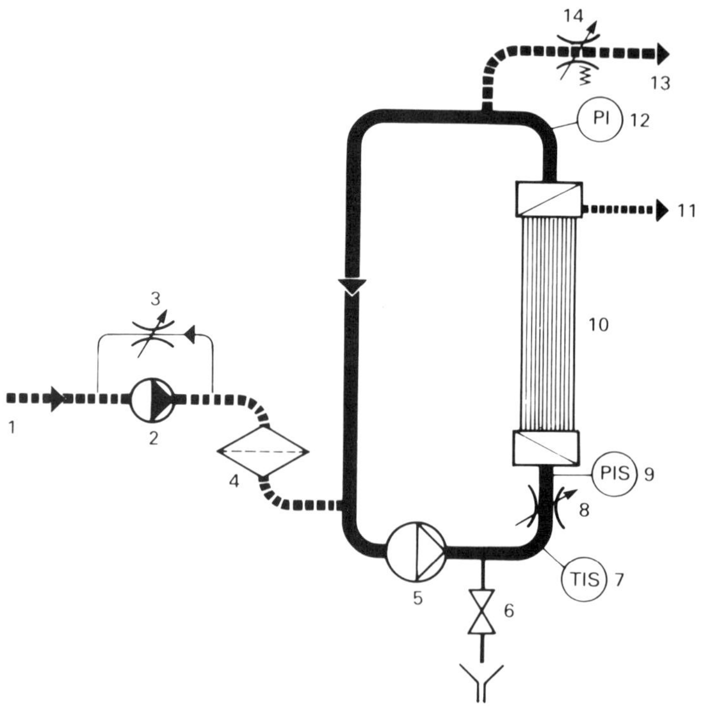

Fig. 5.1 – Ultrafiltration – batch operation. 1. Feed from batch tank. 2. Feed pump. 3. Feed valve. 4. Feed strainer. 5. Circulation pump. 6. Drain valve. 7. Temperature switch and indicator. 8. Inlet pressure control valve. 9. Pressure switch and indicator. 10. Cartridge. 11. Permeate outlet. 12. Pressure indicator. 13. Bleed (concentrate) outlet to batch tank. 14 Bleed valve.

5.3.2 Continuous operation

In continuous operation, product with the desired concentration has to be produced directly from the UF unit itself. By regulating concentrate discharge, while at the same time supplying fresh feed material, a constant concentration level can be maintained in the circulating system and thus in the finished product.

Flux is lower in continuous operation, as the concentration in the circulation system of the unit is higher than in a batch plant. Therefore the required membrane surface is larger than for a batch operating unit with the same capacity. The preferred method is to design a continuous plant as a series of units where the concentrate from one unit is fed to the subsequent unit, a 'feed and bleed' system. With more than five units the membrane surface area approaches that which would have been required for batch plant. At the same time, the advantages of a continuous plant are maintained. However, a relatively sophisticated control system is required.

5.4 ULTRAFILTRATION APPLICATIONS

The applications of ultrafiltration in the pharmaceutical and biochemical industries, can be divided into four fundamental modes of operation, namely:

(i) concentration – removal of solvent from dilute solutions;
(ii) diafiltration – separation of salts and low molecular weight substances;
(iii) product filtration – separation of sludge and high molecular weight substances;
(iv) fractionation – separation of molecules of different size.

5.4.1 Concentration

All UF membranes have the ability to concentrate dissolved macromolecular substances in a solution. In addition to this fundamental ability, different types of UF equipment, have a number of specific characteristics that make them more or less suitable to fulfil a specific duty.

A very important factor, is the ratio between the minimum manageable product volume, and the filter area. Particularly when concentrating microbial products, a short residence time is essential. The filter area must be large in relation to the product volume, particularly for batch operation. At the same time the filter area is limited as the concentrate volume can never be less than the volume of the circulation circuit. The ratio, circulation volume to filter area, affects the treatment time and thus the potential degeneration of a product.

Another very important factor is the power consumption per unit filter area. This figure can vary from about 200 W m^{-2} to several kW m^{-2}. Not only does the relative power consumption determine the pumping energy, it also determines the cooling necessary to maintain a constant, low processing temperature.

To achieve a high yield after completing concentration, the UF system must be easy to drain efficiently. However, concentrated products often have such a high viscosity that a residual product film on the membrane can represent a high loss of valuable product. To increase yield, a very short backflushing of the membranes could be performed. The dilution of the concentrate by the backflushing liquid can be compensated for by a slight overconcentration. An alternative to backflushing for increased yield is to blow sterile air across the membranes.

A typical example of concentration by the use of UF is the concentration and purification of enzymes. Their molecular weight varies from about 13 000 to more than 1 000 000. Enzymes are proteins with a highly specific catalyst effect and with definite pH and temperature optima for maximum activity. Because of their protein structure, they may be inactivated temporarily or permanently by heat and/or chemicals. Processing of enzymes therefore, must be carried out under carefully controlled conditions in order to maximise recovery of activity.

A large range of enzymes, including pectinases, proteases and carbohydrases, are today produced using ultrafiltration.

5.4.2 Diafiltration

During ultrafiltration, dissolved salts and low molecular weight species pass through the membrane with the solvent, which normally is water. In the case where fluid volume becomes unacceptably low as solids in the retentate increase and the desired level of purification has not been reached, additional solvent may be added to achieve further purification in a technique known as diafiltration.

Although it is feasible to dilute the retentate with pure solvent in a batch operation, and to repeat the concentration, this is not as efficient as adding solvent at a rate equal to the permeate flow before the flux has reached an unacceptable level. In this case, the minimum volume of solvent is required.

When a high degree of purity of the product is required, the amount of water that has to be added is much larger than the product volume. Typically, to reduce the salt concentration by a factor of 100, by diafiltration at constant volume, a water amount of 4.6 times the initial product volume has to be added. Diafiltration is often therefore a time-consuming process.

The treatment of difficult substances can be eased if the addition of water is combined with periodic backflushing.

In the case of continuous multistage systems, solvent can be injected in between stages to effect a compromise between the use of the minimum amount of diafiltration solvent and minimising the membrane area (capital cost) required for the plant. Again, the more interstage points used for diafiltration solvent injection, the closer the quantity of solvent required will approach that of the batch system.

Without labouring the theory behind diafiltration, it can generally be concluded that constant volume batch diafiltration is the most efficient process mode and is generally used in situations involving:

(i) products manufactured in small batch operations;
(ii) products which are chemically and microbiologically stable at the processing temperature and thus can withstand a long residence time during diafiltration.

Continuous diafiltration is normally used for processing in situations where:

(i) the product is manufactured on a relatively large scale and the diafiltration process must match upstream or downstream operations, or
(ii) the product is chemically or microbiologically unstable and a short residence time during diafiltration is necessary.

The diafiltration process is supervised either by measurement of the quantity of water added, or by measuring the permeate conductivity.

5.4.3 Product filtration

The cross-flow technique for continuous cleaning of the membrane surface is a prerequisite to carry out ultrafiltration in an economical way. Even with high fluid velocity and turbulent flow conditions a build up of a gel film on the membrane surface will occur.

This gel layer normally acts as a secondary membrane and gives a lower cut-off than the membrane specification. When concentrating, this gel film gives a lower flux than a clean membrane. Usually this fact is accepted and the plant is designed to compensate for the reduced capacity.

In product filtration, when the valuable substances are contained in the permeate, the conditions are totally different. The product has to penetrate the membrane and a secondary layer can retain the valuable product under uncontrolled conditions with lower yield as a result.

Substances with low molecular weights, such as salts, diffuse without difficulty through this gel film. Organic substances in low concentration and with molecular weights of about 200 are sometimes retained even if the nominal cut-off for the membrane is about 50 000. This secondary membrane effect, gives an unacceptably low product yield and in such cases the application of ultrafiltration is often regarded as not feasible.

A sometimes very efficient way to limit the influence of secondary layers is to apply backflushing of the membranes. Other means include high cross-flow velocity, low trans-membrane pressure and careful pretreatment of the broth.

To improve yield, the product filtration can be followed by diafiltration. The concentration of the product in the permeate will, however, suffer.

Product filtration can be continuous or batchwise. As the permeate contains the product, the permeate flow is practically continuous even for a batch process.

Product filtration through ultrafiltration membranes is increasingly being applied in the purification of antibiotics after fermentation and prior to extraction or chemical modification. Ultrafiltration concentrates and removes the cells and cell debris as well as residual macromolecules, suspended particles and colloids from the spent fermentation medium. The clarified permeate, containing the antibiotic practically free from all insoluble and macromolecular components, is then further treated.

Another application of product filtration is the separation of pyrogens and other bacterial residuals from product liquids and nutrient media.

5.4.4 Fractionation

The combination of product filtration, concentration and diafiltration makes it possible to separate certain molecular sizes. The membranes chosen for the first filtering step have a pore size such that the product can pass as permeate. In the next step the pore size is chosen so that the product remains on the concentration side. This operation is simple in principle but is complicated by several factors.

The pore size of ultrafiltration membranes is not uniform – there is in reality a distribution of pore sizes. This fact has to be taken into account to get an acceptable product yield and purity. Ultrafiltration membranes are also graded according to their retention of spherical molecules. Linear molecules with high molecular weight can therefore permeate membranes which have a lower nominal cut-off.

Molecules can form agglomerates and even bigger complexes can be built up. Therefore pretreatment of the feed such as pH stabilisation, temperature control, etc. can be of great importance. The smaller the ratio of the molecular weights to be separated, the greater are the demands concerning membranes and the operational conditions. Normally, the first step is the most difficult as the sludge concentration is rather high and the product may be heterogeneous.

The product filtration step, in which the permeate contains the product, always delivers a continuous flow of filtrate, even when operated batchwise. In this case variations in product flow – high at the beginning, low at the end – are to be anticipated. The final value depends on the concentration of undesired substances in the remaining concentrate.

In the case where the concentration of these undesired substances is finished by diafiltration to yield more product, changes in product concentration occur during this step.

Continuous product filtration eliminates these obstacles, but more control equipment is needed.

5.5 SUMMARY

In conclusion, the advantages of ultrafiltration to the pharmaceutical industry can be summarised as follows:

(1) UF fulfils the dual requirement of concentration and purification simultaneously.
(2) UF involves no phase change, heat treatment or addition of chemicals.

(3) UF reduces the risk of bacterial contamination and simplifies automated cleaning.
(4) Low pressure ultrafiltration systems which operate in the laminar or transition flow regions do not cause shear damage to retentates.
(5) Full draining of UF systems and mild processing conditions increase overall product yield.
(6) pH and ionic strength are not altered during UF processing.
(7) UF is substantially less expensive than evaporation or freeze-drying.

6

Polyethylene glycol precipitation of a microsomal enzyme

M. A. Winkler,[†] A. Wiseman[‡] and A. M. Sadler[†]
† Department of Chemical and Process Engineering and
‡ Department of Biochemistry,
University of Surrey, Guildford, Surrey, UK

6.1 INTRODUCTION

Several different techniques are available for recovering a particular enzyme from the complex mixture with other enzymes and proteins which usually results from the early stages of production, such as cell disruption. In considering these techniques, it is important to distinguish between 'purification' and 'concentration'. An enzyme recovery procedure may effect enzyme concentration, enzyme purification, or, ideally, both. The product of enzyme concentration contains the required enzyme at a high concentration, but together with other active enzymes and proteinaceous or lipid material. The product of enzyme purification contains the required enzyme at a high specific activity, with low levels of other proteins present, but not necessarily at high concentration. While these two effects are not inherently mutually exclusive, it is unusual for both concentration and purification to be obtained in a single recovery technique.

Recovery techniques may conveniently be divided into two groups: techniques separating solution components principally according to size, and techniques utilising other properties of proteins.

6.2 SIZE FRACTIONATION

The techniques considered here separate enzymes and other components according to particle size and, to some extent, particle shape, and are in common use for enzyme separation.

6.2.1 Membrane processes

Solutions are filtered under pressure through a synthetic polymer membrane, which can have pores with a preselected average diameter in the range 0.05 to 0.5 μm, to achieve the concentration of dilute macromolecular solutes. The process operates at constant pH, which is an advantage in processing pH-sensitive materials, requires no chemical reagents, and is rapid, economical and non-destructive. Such ultrafiltration through successive membranes with progressively decreasing average pore-size can be used to effect size fractionation, but as membranes are susceptible to blinding, preliminary treatment of, for example, a cell homogenate is essential.

6.2.2 Gel 'filtration'

Gel 'filtration' is, in fact, a chromatographic process in which high molecular weight components of a solution passing through a 'filtration' medium have a shorter mean residence time than do low molecular weight components. The 'filtration' medium consists of a gel of crosslinked dextran ('Sephadex')† or of crosslinked polyacrylamide ('Biogel'),‡ formed into beads of diameter 80 to 150 μm. The crosslinking is effected so as to give a well defined average pore size in the beads. When a solution passes through a bed of the beads, components larger than the pore size are denied access to the interior of the beads, and so pass more quickly through the bed than do those smaller than the pore size. The latter are able to penetrate the whole bed, and with the larger volume available to them, take longer to pass through the bed. This does not concentrate any particular component, but is useful as a purification stage following previous fractionation steps. In theory, it is possible to separate a solution into fractions containing different size ranges of components by passage through successive beds with progressively changing pore sizes, although this is not commonly practised.

6.2.3 Centrifugation

Sedimentation of solution components under centrifugal force is a long-established technique at analytical, preparative and production scales, giving fractionation of components according to size, density and, to some extent, shape. If the sedimentation follows Stokes' Law, then the time 't' taken for a single spherical particle of diameter 'd' to settle from a distance z_1 to z_2 from the liquid surface in a radial direction is given by

$$t = \frac{18\eta}{d^2 . \omega^2 . \Delta\rho} \cdot \ln \frac{L + z_1}{L + z_2} \qquad (1)$$

where 'L' is the radius of the inner liquid surface, 'ω' the angular velocity, '$\Delta\rho$' the density difference between the particle and the liquid, and 'η' the liquid viscosity (Herdan 1960).

† From Pharmacia, Sweden.

‡ From Biorad, USA.

From this relation, it is clear that sedimentation of very small particles requires centrifugation at very high speed of rotation and/or for long periods of time. With organic materials, this problem is exacerbated by their densities being, in general, little different from those of the aqueous media in which they are processed. Centrifugation is thus a suitable technique for most purposes of enzyme recovery only when the effective particle size of the required enzyme is increased, by aggregation or by attachment to a larger entity.

6.3 OUTLINE OF OTHER FRACTIONATION TECHNIQUES

The nature of the techniques outlined here may in most cases be generalised as utilising the electric charge of proteins and/or the ability of proteins to attach to other entities.

6.3.1 Affinity chromatography

This technique is arguably the ultimate in selectivity for enzyme purification, as it utilises the highly specific interaction, termed 'molecular recognition', between an enzyme and its substrate, or between an enzyme and the protein antibody raised against its protein component in a suitable animal. The substrate, or the antibody, is attached to a suitable solid support medium which is mixed with a solution containing the required enzyme. The enzyme attaches preferentially to its substrate or its antibody, and is subsequently eluted with a solution of different pH or ionic strength. The technique is most useful as the final purification stage in enzyme recovery. Some contamination is, however, likely to occur, as a result of non-specific binding of other materials, to the affinity matrix or to the bound enzyme molecules.

6.3.2 Ion exchange

Enzymes in solution carry a net electric charge, which depends on the pH of the surrounding medium, and are thus amenable to recovery by ion exchange techniques. At its 'iso-electric pH' ('IEP'), the net electric charge on a protein is zero. At pH values below the IEP, a protein has a net positive charge, and above the IEP a net negative charge. An appropriate ion exchange medium for collecting a particular enzyme can then be selected according to the iso-electric point of the required enzyme and the pH of the solution containing it. The enzyme is subsequently eluted from the ion exchange medium with a buffer solution of a different pH and/or ionic strength, as was noted above for affinity chromatography. The process is not highly specific, as the ion exchange medium collects all proteins with a charge similar to that of the required enzyme. Where an enzyme is stable at its iso-electric point, fractionation can be effected by using solutions with pH successively above and below the IEP of the enzyme, using anion and cation exchange media successively. In addition to the usual ion exchange media described elsewhere in this volume (see Chapter 7), some biological applications require cellulose or dextran that has been 'derivatised' with compounds containing ionisable functional groups.

6.3.3 Electrophoresis

The net electric charge on an enzyme, together with its molecular size and, to some extent, shape, can be utilised in separating it from a mixture of proteins by electrophoresis. The dissolved mixture, held in a viscous medium at a suitable pH, is placed in an electrostatic field. The components of the mixture then slowly separate in the electric field, because highly charged components move faster than components with a low net charge. In addition, small particles move faster than large particles through the viscous medium employed. The simplest form of this technique has been used for many years in clinical biochemistry laboratories for identifying components of complex protein mixtures, such as those in blood plasma. These form recognisable patterns, like spectra, in which unusual components, or missing substances are fairly readily identified. The technique is also available in scaled-up versions suitable for preparative purposes. In continuous flow versions, a stream of the protein solution passes between a pair of long electrodes, and components migrate towards either electrode according to charge and size, so that the output stream can be divided physically into fractions with differing compositions. Separation is impaired by turbulence in the process stream, so it is important to maintain laminar flow conditions in the liquid. Turbulence can arise from convection currents resulting from local heating effects, for example. An interesting means of obviating this is to carry out the separation in conditions of zero gravity, in an orbiting space-shuttle. According to a report in *New Scientist* (29 November 1984), a secret pharmaceutical product, believed to be a hormone, was recovered in orbit with, it was claimed, a fourfold improvement in yield and a 700 times increase in throughput with continuous flow electrophoresis compared with a similar process on earth. In the 'Biostream' process developed at Harwell (marketed by CJB Ltd, Portsmouth, UK), the process solution flows upwards between vertical annular electrodes, laminar flow being stabilised by rotation of the outer electrode. Materials ranging from small molecules to whole cells can be separated in free-flowing solution at rates claimed to be typically up to $100\,g\,h^{-1}$ of protein (Wilkinson 1984). Other related techniques include iso-electric focusing, which uses special gel supports and buffers to concentrate a protein at its iso-electric point (section 6.3.2).

6.3.4 Foam fractionation

Proteins are surface-active substances and can be separated according to their differences in surface activity by foam fractionation. When gas bubbles are foamed in a solution containing proteins, certain proteins are preferentially adsorbed at the gas–liquid interface and are carried up by the bubbles to appear in the resulting foam layer on top of the liquid. The method was successful in effecting the first separation of rennin and pepsin, by Andrews and Schütz in 1945. These proteolytic enzymes had been so difficult to separate that, for a long time, they were thought to be identical. Foam fractionation is not, however, widely used for enzyme recovery, due partly to the belief that it causes protein denaturation. It has been shown that, with a suitable choice of operating conditions, a wide range of enzymes can be foam fractionated with no loss of activity (Thomas & Winkler 1977).

6.3.5 Partition

Certain solutions of non-ionic polymers can separate out to form two distinct aqueous phases. Proteins and other components of the solution will be distributed unequally between the two phases, as in solvent extraction. Phase separation occurs when the solution contains two incompatible polymers, such as dextran and polyethylene glycol, or an electrolyte and an incompatible polymer, such as potassium phosphate and polyethylene glycol (Kroner *et al.* 1978). The separation achieved depends on several factors, including the types of polymer and their concentrations and average molecular weights, the ions present and the ionic strength, pH and temperature of the solution. Thus a considerable degree of purification of a particular enzyme can be obtained using successive partitions with different sets of conditions. The process is suitable for large-scale enzyme isolation, including enzyme recovery from crude cell preparations using commercial centrifuges. The principal disadvantage of the process is the contamination of the product with polymer, which must be removed by dialysis or other membrane processes in a subsequent step, and concentration may also be necessary.

6.4 PROTEIN AGGREGATION

From section 6.2.3, it is clear that, if sedimentation is to be employed in enzyme recovery, the enzymes need to be in the form of particles several orders of magnitude larger than the individual molecules for sedimentation to be of realistically short duration.

The process of bringing together molecules to form large, sedimentable particles is known by a number of different names – 'aggregation', 'agglomeration', 'coagulation', 'flocculation' or simply 'precipitation' – which tend to be employed indiscriminately by most authors. The essence of the process is that a solution or suspension is first destabilised, then the destabilised components are induced to come together and form agglomerates (Bratby 1980). These effects are brought about on an industrial scale electrostatically with salting-out, metal-ion, isoelectric and 'organic' solvent precipitation, and non-electrostatically with non-ionic polymers. The agglomerates are then recovered by sedimentation, usually by centrifugation, but in some cases by gravity settling.

6.4.1 Salting-out

The solubility of proteins in solutions of neutral salts varies with the ionic strength of the solution. At low values of ionic strength, protein solubility increases with increasing ionic strength, an effect known as 'salting-in'. As the ionic strength increases further, however, protein solubility reaches a maximum and then decreases.

The decrease of solubility with increasing ionic strength at higher values of ionic strength is called 'salting out' and follows an exponential decay pattern, described empirically by

$$\ln S = \ln S_0 - k_s \, . \, J \tag{2}$$

where 'S' is the protein solubility in a solution of ionic strength 'J', 'S_0' is a hypothetical (not actual) protein solubility in the solution at zero ionic strength and 'k_s' a 'salting-out coefficient'. While 'S_0' is strongly dependent on the nature of the protein, 'k_s' is affected only slightly by the nature of the protein and depends mostly on the nature of the salt in the solution (Cohn & Edsell 1943). Thus salting-out is particularly suitable for selective precipitation of proteins, and is widely used on both laboratory and industrial scales. Its limitation is that proteins are susceptible to denaturation in solutions of very high ionic strength. Ammonium sulphate is commonly used for salting-out, as it has a high solubility and is cheap, but residual salt in the final product can cause toxicity in clinical preparations and impart a detectable off-taste to food products. The spent ammonium sulphate solution may give rise to disposal problems, as it is highly polluting, having a high biological oxygen demand, and is toxic to aquatic life at low concentrations (Winkler 1981).

6.4.2 Metal-ion precipitation

The solubility of a protein is reduced when its exposed anionic groups combine with metal ions in the surrounding solution. The effect is particularly marked with polyvalent metal ions, and is further enhanced if the dielectric constant of the solution is reduced, by the addition of low concentrations of, for example, ethanol to the solution.

6.4.3 Iso-electric precipitation

Proteins in general have minimum solubility at their iso-electric point, the pH of the solution in which the protein has a net zero electric charge. Thus the effects of other precipitation techniques are generally considerably enhanced if carried out at the iso-electric pH of the required protein. It should, however, be borne in mind that some proteins are readily denatured at their iso-electric pH.

6.4.4 Precipitation with 'organic' solvents

Reducing the dielectric constant of a solution by addition of organic solvents, such as methanol, ethanol or acetone, is a very effective method of decreasing protein solubility. The resultant increase in intermolecular electrostatic effects encourages protein aggregation, and thus precipitation. Similarly, organic solvents with relatively high dielectric constants, such as dichloroacetic acid, are good protein solvents.

Low dielectric solvent precipitation can be used in conjunction with metal-ion precipitation or with salting-out, and protein fractionation can also be effected by balancing solvent precipitation with salting-in at low ionic strengths.

The solvents reduce the density of the liquid phase, thus increasing the density difference between the liquid and the protein agglomerates and so assisting sedimentation in accordance with eqn (1).

There are, however, a number of drawbacks to the use of organic solvent precipitation. The solvents are flammable, and strict safety precautions are necessary in plant design and operation, and in materials handling and storage. Solvent recovery and concomitant losses are expensive, and proteins are readily denatured by organic solvents.

6.4.5 Non-ionic polymers

Non-ionic polymers do not form ions in solution and thus do not interact electrostatically with solute proteins. Proteins are forced to aggregate, and thence precipitate, as a result of steric exclusion from the solution by the non-ionic polymer. This mechanism was proposed by Laurent (1963) from investigation of the effect of dextran on protein solubility. Edmond and Ogston (1968) suggested that the increase in chemical potential of one solute in the presence of another leads to its value exceeding that of the corresponding solid phase. If this occurs within the region of the phase diagram that is stable with respect to the separation of liquid phases, then precipitation of solid phases will occur before liquid phase separation. This leads to an expression very similar in form to the Cohn salting-out equation (eqn (2)):

$$\ln S = K_G - B.G \tag{3}$$

where 'S' is the concentration of protein in a solution containing a concentration 'G' of non-ionic polymer, and 'K_G' and 'B' are coefficients characteristic of the protein and the polymer. Juckes (1971) developed a similar expression from the Cohn salting-out equation (eqn (2)), based on the assumption of a steric exclusion mechanism, which gave a good description of the precipitation of brome grass virus, haemoglobin, ovalbumin and bovine serum albumin with polyethylene glycol (PEG). PEG precipitation of plasma protein was modelled semi-empirically by Hasko *et al.* (1982) in terms of plasma protein solubility as a function of PEG concentration, expressed in the form

$$\ln(S - S^*) = K^* + b.\ln(1 - G) \tag{4}$$

for comparison with eqns (2) and (3), where 'S' is the protein concentration in a solution containing a concentration 'G' of PEG, 'S^*' is the final residual protein concentration unaffected by further increases in 'G', and 'K^*' and 'b' are characteristic coefficients. A model of PEG precipitation of a microsomal enzyme, based on particle statistics, has been proposed (Sadler *et al.* 1985) and is discussed in detail further on (section 6.5.3).

The suitability for protein precipitation of several high molecular weight non-ionic polymers, polyethylene glycol, dextran, nonyl phenyl ethoxylate, polyvinyl alcohol and polyvinyl pyrrolidone, were compared by Polson *et. al.* (1964). Polyethylene glycol was found to be the most suitable of these, as its solutions were less viscous and its use caused virtually no protein denaturation at room temperature. From an investigation of the effect of PEG concentration on the solubility of bovine serum albumin, using PEG of different average molecular weights, 1000, 7500, 19 000, and 100 000, Knoll & Hermans (1983) concluded that the steric exclusion mechanism accounted satisfactorily for protein precipitation by PEG.

6.5 MICROSOMAL ENZYMES

Particular enzymes in eukaryotic cells are bound to an intracellular membrane known as the 'endoplasmic reticulum'. To recover such enzymes from biomass, the cells must first be disrupted, and during disruption, the endoplasmic reticulum is pinched off to form small vesicles. These vesicles are commonly known as 'microsomes' or, when recovered by high-speed centrifugation, as a 'microsomal fraction'. Such a vesicle is several orders of magnitude larger than the individual enzyme molecule, and comprises a range of enzymes and other proteins, as well as the original membrane fragment to which they are bound. This is well illustrated by the enzyme studied by the authors, cytochrome P-450 of *Saccharomyces cerevisiae* (brewer's yeast), a microsomal fraction preparation of which contains only about 0.1 to 0.2% by weight of cytochrome P-450.

Most of the enzyme recovery techniques considered in the previous sections are appropriate only to enzymes in solution. To use them in the separation of a microsomal enzyme, the enzyme must first be detached from its membrane fragment and brought into solution. This is effected by 'solubilisation' with a surface-active substance, such as sodium cholate. Solubilisation immediately introduces a problem, however. When removed from its microenvironment in the microsome, a microsomal enzyme becomes very much more susceptible to denaturation. It is therefore preferable to carry out concentration procedures as far as possible with the enzyme in its microsomal form, then solubilizing at a late stage for final purification. Unfortunately, this procedure also has its problems. Although large compared to individual enzyme molecules, microsomes are nevertheless small for the purposes of sedimentation, with a characteristic dimension typically less than 0.1μm. Recovery by sedimentation thus requires centrifugation at high speeds and/or for long periods of time (eqn (1)). The intracellular localisation of an enzyme can, however, simplify a separation process by confining the enzyme to a particular fraction defined by the size and/or shape of the organelle. Thus the standard procedure for sedimentation of microsomal cytochrome P-450 from a microbial cell disruptate, for example, requires centrifugation for 1 hour with a relative centrifugal 'force' (RCF) of $10^5 \times g$. While this is perfectly feasible in bench-scale preparation, the low volume capacity of high-speed centrifuges gives rise to a serious bottleneck in the processing of more than a few hundred millilitres of cell disruptate.

6.5.1 Cytochromes P-450

Cytochrome P-450 enzymes are involved in many cellular biosynthetic and metabolic reactions in animals, microorganisms and plants. Several of these enzymes are targets for inhibition by a number of chemotherapeutic and agrochemical compounds (Coulson, King & Wiseman 1984). Particular strains of brewer's yeast, *Saccharomyces cerevisiae*, produce a cytochrome P-450 enzyme when grown under conditions of glucose repression (Wiseman & Lim 1975), and, after careful optimisation of yeast growth conditions (Salihon 1984; Salihon *et al.* 1983, 1985), a specific enzyme yield of 126 nanomoles per gram (dry weight) of yeast can be obtained. Fermentation thus provides a potentially cheap, large-scale source of microbial cytochrome P-450 for investigation of these inhibition effects. Cytochrome P-450 of *S. cerevisiae* is also of particular interest in displaying hydroxylase activity on known carcinogens such as benzo(*a*)-pyrene, both *in vitro* and *in vivo.* Attempts to precipitate cytochrome P-450 from centrifuged yeast disruptate with ammonium sulphate, calcium ions or acetone resulted in extensive enzyme loss, due presumably to its sensitivity to conditions of high ionic strength and low dielectric constant. It should, however, be mentioned that ammonium sulphate has been used successfully in precipitating cytochrome P-450 from a microsomal preparation solubilised with sodium cholate (Azari & Wiseman 1982), and calcium-ion precipitation was used successfully in separating microsomal cytochrome P-450 from another yeast, *Candida tropicalis* (Kappeli *et al.* 1982) using multi-stage centrifugation. A preparation of cytochrome P-450 was required for immobilisation, in the development of an immobilised-enzyme reactor. For this purpose, concentration of the enzyme is more important than its purification.

6.5.2 Polyethylene glycol precipitation of microsomal cytochrome P-450

Yeast microsomal cytochrome P-450 is sensitive to separation by conventional techniques of salting-out, metal-ion and 'organic' solvent precipitation. The general technique of precipitation with polyethylene glycol has been reported (Juckes 1971) as a gentle process giving high recovery level of different proteins. A preliminary study on the precipitation of cytochrome P-450 from disrupted yeast biomass using polyethylene glycol with an average molecular weight of 6000 (PEG 6000) (Sadler *et al.* 1983) showed that about 90% enzyme recovery could be obtained with a low concentration of PEG 6000, about 6%, and centrifugation with a RCF of only 2300 × g for 20 minutes. Some imprecise fractionation of the enzyme also occurred (Fig. 6.1), but, as mentioned in section 6.5.1 enzyme purification was of secondary importance to enzyme concentration. Further studies using centrifugation at different RCF values, as well as with different PEG 6000 concentrations were carried out, but with a constant centrifugation time (Sadler *et al.* 1985). These indicated that the PEG effectively 'replaces' RCF in achieving a given level of enzyme recovery (Fig. 6.2). This enables a high recovery of the microsomal enzyme to be obtained from disrupted yeast biomass with minimal denaturation, using PEG 6000 concentrations of only a few per cent, combined with centrifugation using RCF values of only a few thousand × g, to give a high concentration, low purity enzyme preparation, suitable for immobilisation.

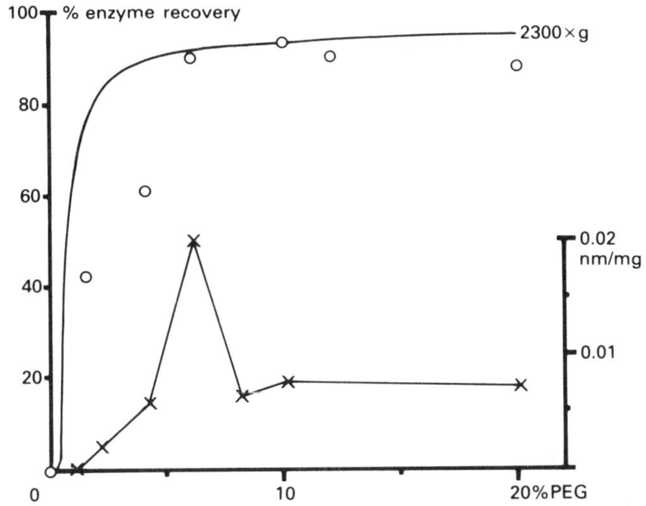

Fig. 6.1 – Precipitation profiles of microsomal cytochrome P-450 from *S. cerevisiae* with different PEG concentrations and centrifuging for 20 min at 2300 × g. ○ – recovery of cytochrome P-450 as a percentage of that in the centrifuged disruptate. × – Specific concentration of enzyme in nmol cytochrome P-450 per mg protein in sedimented PEG precipitate. (Reproduced from Sadler *et al.* (1985) by permission of *Biochem. Eng. J.*)

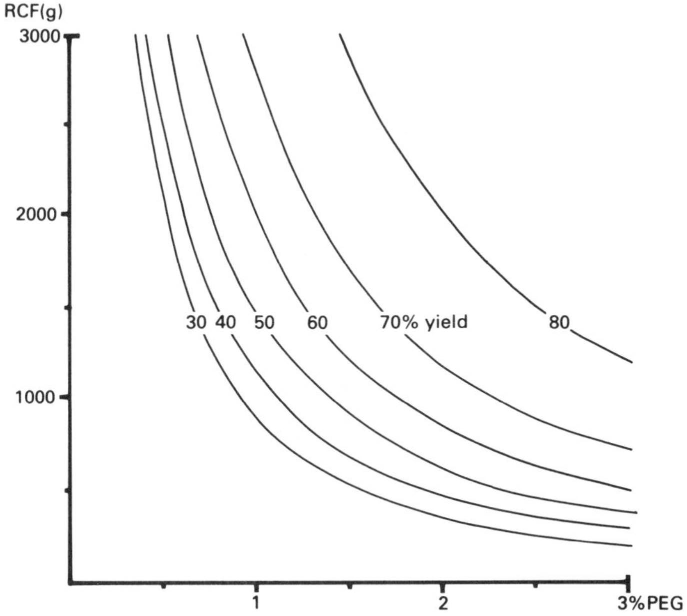

Fig. 6.2 – Contour plot illustrating theoretical yields of microsomal cytochrome P-450, as percentages of that in the centrifuged disruptate, obtainable with different combinations of RCF value and PEG concentration. (Reproduced from Sadler *et al.* (1985) by permission of *Biochem. Eng. J.*)

6.5.3 Particle statistics model of PEG precipitation

Enzyme recovery clearly increases with both increasing PEG concentration and increasing RCF used, but not linearly in either case, and a given level of enzyme recovery can be obtained using several different combinations of PEG concentration and centrifuge RCF. A quantitative model of the effects of PEG 6000 concentration and centrifuge RCF on the recovery of cytochrome P-450 from disrupted yeast biomass was advanced by Sadler *et al.* (1985). This was based on the assumptions that the PEG concentration used affects both the average size of protein agglomerates formed and their size distribution. As increasing concentrations of PEG are used, increasing proportions of the protein present are forced out of solution. The protein then agglomerates, forming particles whose average size and range of size distribution increases with increasing PEG concentration used.

If a given PEG concentration produces a definite agglomeration pattern, and aliquots of such a suspension are centrifuged for a constant time at different RCF values, then the different protein recoveries obtained should follow the size distribution pattern of protein agglomerates formed with that PEG concentration. From eqn (1), for a given system with constant centrifuge configuration and centrifugation time,

$$d^2 . \omega^2 = \text{constant} \tag{5}$$

and since the RCF value is proportional to ω^2, then

$$d \propto R^{-1/2} \tag{6}$$

where 'R' is the RCF value. This means that the diameter of the largest particle that can escape sedimentation is inversely proportional to the square root of the RCF used, and a plot of the unrecovered enzyme fraction against the reciprocal square root of the RCF used should follow the size distribution pattern of the protein agglomerates. The pattern was likely to follow a 'Normal' distribution, as materials showing a 'Normal' particle size distribution are found chiefly among the particulate substances produced by chemical processes like precipitation (Herdan 1960).

Assuming 'Normal' particle size-distribution, the percentage enzyme recovery, 'P' is then given by

$$P = 100\{1 - \phi(h)\} \tag{7}$$

where

$$h = \frac{1}{\sigma}(R^{-1/2} - R_{50}^{-1/2}) \tag{8}$$

'R_{50}' is the RCF giving 50% enzyme recovery, '$\phi(h)$' is the cumulative normal distribution function of 'h' and 'σ' its standard deviation.

The RCF values giving 50% enzyme recovery were correlated logarithmically with 'G', the concentration of PEG 6000 used, to give

$$R_{50} = A \,.\, G^{m} \tag{9}$$

with results indicating that 'm' is about −1.24. Preliminary probability plots also indicated that the standard deviations were in the range 0.5 to 0.7 × $R_{50}^{-1/2}$ for each PEG concentration, and a value of 'σ' of 0.53 × $R_{50}^{-1/2}$ gave a reasonable overall fit for the PEG concentrations used. This implies that the spread of agglomerate sizes increases with increasing PEG concentration and thus with average agglomerate particle size, so that

$$\sigma = 0.53(A^{-1/2}) \,.\, G^{0.62} \tag{10}$$

Plots of percentage enzyme recovery for different PEG concentrations and RCF values can now be constructed, and can be seen to give a reasonable description of the experimental results (Fig. 6.3).

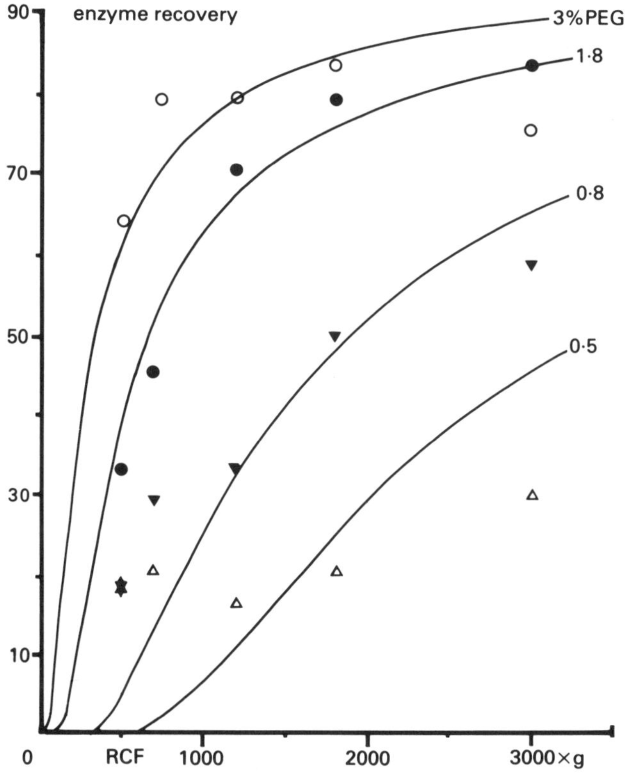

Fig. 6.3 – Precipitation profiles of microsomal cytochrome P-450 from *S. cerevisiae* with different PEG concentrations and centrifuging for 5 min with different RCF values. Recovery of cytochrome P-450, as a percentage of that in the centrifuged disruptate, using PEG concentrations of: ○ – 3%; ● – 1.8%; ▲ – 0.8%; △ – 0.5%; (Reproduced from Sadler *et al.* (1985) by permission of *Biochem. Eng. J.*)

6.6 NOMENCLATURE

A	RCF giving 50% enzyme recovery with a 1% PEG concentration
B	characteristic coefficient in non-ionic polymer precipitation
b	characteristic coefficient in PEG precipitation
d	diameter of a spherical particle
G	concentration of non-ionic polymer
g	acceleration due to gravity
h	parameter defined in eqn (8)
IEP	iso-electric point, iso-electric pH
J	ionic strength
K_G	characteristic coefficient in non-ionic polymer precipitation
K^*	characteristic coefficient in PEG precipitation
k_s	salting-out coefficient
L	radius of inner liquid surface in a centrifuge
m	empirical exponent
P	percentage protein recovery
PEG	polyethylene glycol
PEG 6000	PEG with an average molecular weight of 6000
R	relative centrifugal force
R_{50}	relative centrifugal force giving 50% enzyme recovery
RCF	relative centrifugal force
S	protein concentration or solubility
S_0	hypothetical protein solubility in a solution of zero ionic strength
S^*	final residual protein concentration
t	time taken for sedimentation
z_1, z_2	radial distances from inner liquid surface in centrifuge
$\Delta\rho$	density difference between particle and liquid
η	liquid viscosity
$\phi(h)$	cumulative normal distribution function of 'h'
σ	standard deviation of $\phi(h)$
ω	angular velocity

REFERENCES

Azari, M. R. & Wiseman, A. (1982) Purification and characterization of the cytochrome P-448 component of a benzo(*a*)pyrene hydroxylase from *Saccharomyces cerevisiae*. *Anal Biochem.* **122**, 129–138.

Bratby, J. (1980) *Coagulation and Flocculation with an Emphasis on Water and Wastewater Treatment.* Croydon, Uplands Press.

Cohn, E. & Edsall, J. T. (1943) *Proteins, Amino Acids and Peptides.* New York, Academic Press.

Coulson, C. J., King, D. J. & Wiseman, A. (1984) Chemotherapeutic and agrochemical applications of cytochrome P-450 and Ligands. *Trends in Biochemical Sciences* October, 446–449.

Edmond, E. & Ogston, A. G. (1968) An approach to the study of phase separation in ternary aqueous systems. *Biochem. J.* **109**, 569–576.

Hasko, F., Vaszileva, R. & Halasz, L. (1982) Solubility of plasma proteins in the presence of polyethylene glycol. *Biotechnol. Bioeng.* **24**, 1931–1939.

Herdan, G. (1960) *Small Particle Statistics.* 2nd edn. London, Butterworths.

Juckes, I. R. M. (1971) Fractionantion of proteins and viruses with polyethylene glycol. *Biochim. Biophys. Acta* **229**, 535–546.

Kapelli, O., Sauer, M. & Fiechter, A. (1982) Convenient procedure for the isolation of highly enriched cytochrome P-450 containing microsomal fraction from *Candida tropicalis. Anal. Biochem.* **126**, 179–182.

Knoll, D. & Hermans, J. (1983) Polymer-protein interactions. *J. Biol. Chem.* **258(a)**, 5710–5715.

Kroner, K. H., Hustedt, H., Granda, S. & Kula, M. R. (1978) Technical aspects of separation using an aqueous two-phase system in enzyme isolation processes. *Biotechnol. Bioeng.* **20**, 1967–1988.

Laurent, T. C. (1963) The interaction between polysaccharides and other macromolecules. *Biochem. J.* **89**, 253–257.

Polson, A., Potgeiter, G. M., Largier, J. F., Mears, G. E. F. & Joubert, F. J. (1964) The fractionation of protein mixtures using linear polymers of high molecular weight. *Biochim. Biophys. Acta* **104**, 463–475.

Sadler, A. M., Winkler, M. A. & Wiseman, A. (1983) Recovery of cytochrome P-450 from *Saccharomyces cerevisiae* using polyethylene glycol precipitation. *Biochem. Soc. Trans.* **11**, 402.

Sadler, A. M., Winkler, M. A. & Wiseman, A. (1985) Recovery of cytochrome P-450 from yeast using low-speed centrifugation. *Biochem. Eng. J.* **30(3)**, B43–49.

Salihon, J. (1984) Optimization of yield of an intracellular enzyme from yeast. Ph.D. Thesis, University of Surrey.

Salihon, J., Winkler, M. A. & Wiseman, A. (1983) Optimization of culture conditions for yield of cytochrome P-450 in *Saccharomycos cerevisiae. Biochem. Soc. Trans.* **11**, 401.

Salihon, J., Winkler, M. A. & Wiseman, A. (1985) Microprocessor controlled stirred tank fermentation in optimization of yield of cytochrome P-450 in *Saccharomyces cerevisiae* NCYC 754. *Biochem. Soc. Trans.* **13**, 526.

Thomas, A. & Winkler, M. A. (1977) Foam separation of biological materials. In: Wiseman, A. (ed.), *Topics in Enzyme and Fermentation Biotechnology*, Vol. 1. Chichester, Ellis Horwood, pp. 43–71.

Wilkinson, J. A. E. (1984) Personal communication (CJB Developments Ltd).

Winkler, M. A. (1981) *Biological Treatment of Waste-water.* Chichester, Ellis Horwood.

Wiseman, A. & Lim, T. K. (1975) *Biochem. Soc. Trans.* **3**, 974–977.

7

Ion exchange resins and adsorbents in the chromatographic extraction of antibiotics

M. Pirotta
Rohm and Haas Company, European Region,
Via Vittor Pisani 26, 20124 Milan, Italy

7.1 INTRODUCTION

Ion exchange resins are insoluble electrolytes that behave as if the ionic groups were soluble. They consist of a matrix and a functional group. From a functional group point of view, the performance of an ion exchange resin is significantly influenced by the presence of organic and inorganic ions, their concentration, their dissociation constants and their molar ratios.

7.2 KEY FACTORS FOR THE CHOICE OF AN ION EXCHANGE RESIN EXTRACTION SYSTEM

The following factors must be evaluated in order to decide whether an ion exchange technique can be used for a specific extraction: the chemical group of the resin, i.e. weak or strong acid; weak or strong base (resin effect); the ionic form of the resin (selectivity effect); the pH of the solution and how it is obtained (ionisation effect).

The factors which have to be considered in order to evaluate whether an ion exchange system is economically attractive are: the best possible ionic composition of the solution to be treated; the best hydraulic conditions for the column run; an engineering system which, during pilot experiments, permits extrapolation of the results obtained to an industrial plant; reduction of the mixing of each process phase; optimisation of flow rates of the different steps of the process, temperature, pressure, viscosity, density, pH of the displacement solution, pH of the eluate, water quality for making up process solutions, resin bed depth.

Although the importance of these factors is known to people skilled in the art, the reason for their importance is not always obvious.

The ion exchange groups in ion exchange resin (IER) beads are not all equally accessible to the ionic substrate. The kinetics of the exchange is controlled by diffusion factors (Fig. 7.1). In the diffusion layer adhering to the bead, there is a gradient of concentration of both A and B. The kinetics of IER is controlled by either the rate of ion diffusion within the resin bead, or in the laminar film around it (diffusion layer). The rate of ion exchange is determined by the slower of the two kinetic rates.

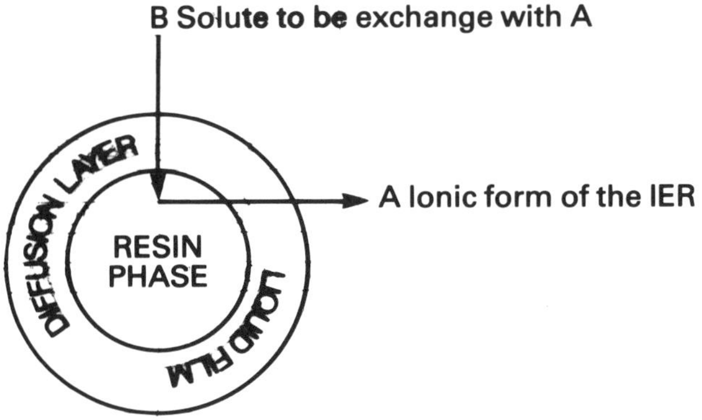

Fig. 7.1 – Ion exchange resin kinetics.

If the hydrated ions that react with the ionic moiety of the resin are significantly bulky, some hindrance could occur and the 'total working capacity' obtained from the resin would be lower than expected. If the hydrated ions are sufficiently small to easily penetrate the resin beads, and the contact time is sufficiently long, the 'working capacity' would approach the total ion exchange capacity of the resin, which is reached only at the equilibrium conditions.

Since ion exchange on these polymers follows the Gulbert and Waage law, the concentration and the activity coefficient of the ions, etc. are also factors that one must not forget. The affinity of the resin towards a certain ionic substrate is, of course, influenced by the type and ionic form of the functional group of the exchanger, but, particularly in the case of antibiotics, it is also significantly influenced by the porosity of the three-dimensional matrix to which this group is attached.

Different ionic forms of the same resin cause changes in the pore size of the matrix and, consequently, its kinetics is affected. Ion exchange takes place through the water coordinated to the ionic group of the resin and, consequently, the kinetics of exchange is also a function of the water solubility of the ionic form in which the substrate is present in the solution. The more dilute the substrate, the more time is required for the ion exchange reaction to reach the equilibrium. When the exchange reaction time equals the equilibrium reaction time, no further effective capacity can be obtained from the system.

Space velocity (h^{-1}) has an important effect on the selectivity and operating capacity of an ion exchange resin, particularly in systems such as fermentation broths. During the optimisation phase of an extraction process, it is advisable to investigate the range of optimal space velocities and choose that most suitable for the process and hydraulic conditions of the particular system.

Since diffusion coefficients and solute concentrations can significantly influence the performance of an ion exchange resin, one should take all possible precautions to prevent heterogeneity of the resin bed, air bubbles, channelling, resin floating, etc. and mixing of solutions used in the different stages of the extraction process.

It is stated in the literature on water treatment that an ion exchange bed should be deeper than 60 cm in order to avoid the formation of channels in it. This is not always true for pharmaceutical extractions. The main reason is that water treatment operations generally use much higher linear and space velocities. For pharmaceuticals one should consider the smallest bed depth satisfactorily used in the laboratory as the minimum to be used for the plant. If, for example, a good chromatographic separation is obtained using a 40 cm bed depth there are no kinetic reasons why comparable results should not be obtained in an industrial plant provided an equivalent distribution and collection system is used. However, other reasons such as hydraulic, mechanical or contingent considerations may lead to the construction of an ion exchange resin column where the resin has a deeper bed depth.

7.3 CARBOXYLIC IER USED AS POLAR ADSORBENTS

Weak electrolyte resins are dissociated only in certain pH ranges. This phenomenon has been industrially exploited for separation of proteinaceous materials, using carboxylic ion exchange resins. At acidic conditions below pH 4.5, methacrylic and acrylic ion exchangers are undissociated and the ion exchange resin becomes a polar adsorbent, which can perform separations by means of hydrogen bonding without ionic binding with the substrates. Several difficult chromatographic separations have been accomplished using this technique (Fig. 7.2).

7.4 INFLUENCE AND IMPORTANCE OF THE MATRIX

7.4.1 Chemical considerations

There are various types of matrix, such as, for example, styrene–DVB and acrylate–DVB, etc. For the sake of illustration we shall consider here only the styrene–DVB types, which are those most commonly used and particularly the synthesis of the strong base types.

A strong base ion exchange resin is synthesised, as follows, by copolymerisation of styrene and DVB (Fig. 7.3). The % DVB bridges influence the distance between the styrene chains and, consequently, the porosity of the gel phase. This step of the synthesis and the related concept of the gel porosity is common to all ion exchange resins and adsorbents based on styrene–DVB copolymerisation.

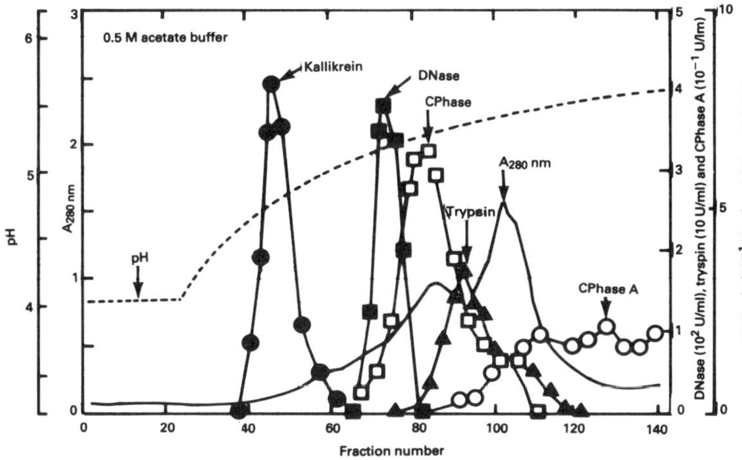

Fig. 7.2 – Example of polar hydrophobic absorption (carboxylic adsorbents). The adsorption-elution mechanism of proteins on Amberlite CG-50 may be as follows. At pH 4.5, proteins are adsorbed by hydrophobic affinity on loci of the resin, each of which includes at least one carboxyl group. As the pH is increased, the carboxyl groups dissociate, resulting in a decrease of the hydrophobicity and an increase of the negative charge of the loci; thus, the proteins are now retained by the remaining hydrophobic affinity *plus* the increased electrostatic affinity or *minus* the increased electrostatic repulsion. (Sasaki *et al.* 1979).

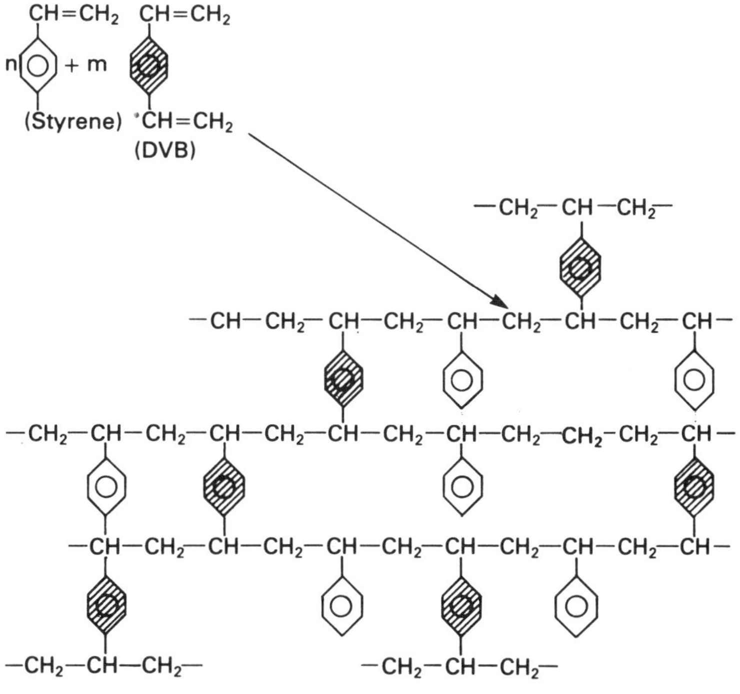

Fig. 7.3 – IER copolymerisation.

To produce a strong acid resin, the copolymer in Fig. 7.3 is sulphonated. For anionic resins $-CH_2Cl$ is introduced in the *para* position of the benzene ring prior to forming a quaternary ammonium group (Figs. 7.4 and 7.5).

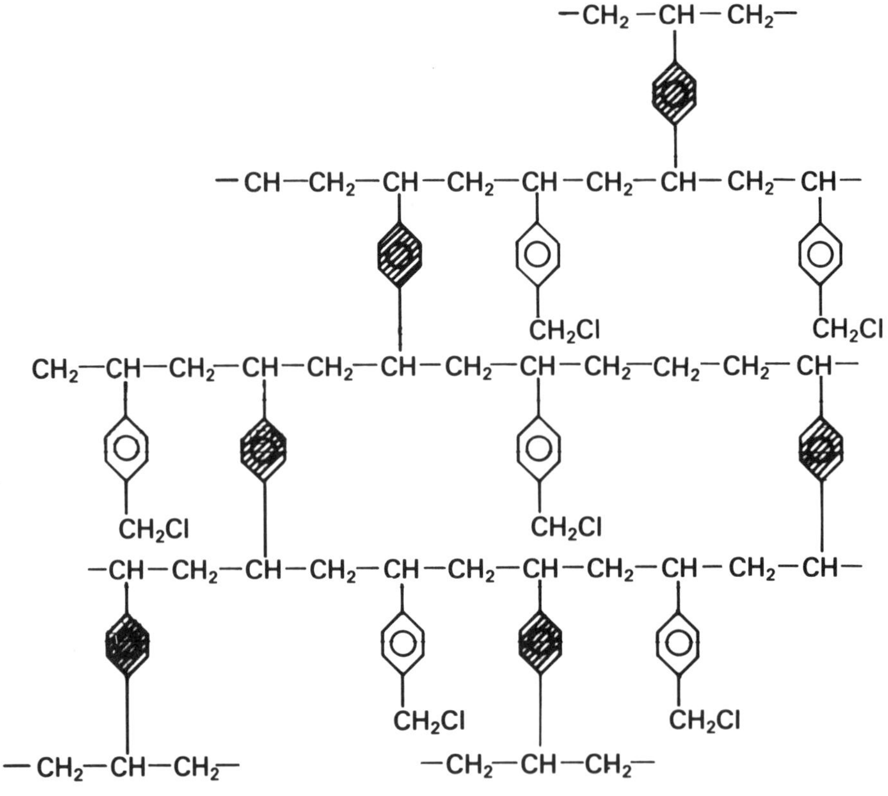

Fig. 7.4 – Strong base anionic resin synthesis.

From Fig. 7.4, it can be seen that the $-CH_2Cl$ groups can inter-react forming what is called secondary crosslinking. The side reaction increases the physical stability of the polymer and reduces the gel phase porosity. The extent of such a side reaction can be controlled by using appropriate catalysts. As a result, a resin having a broad range of properties can be made from a certain copolymer composition by changing the chloromethylation conditions. Further, resins having apparently identical physical characteristics can be made from copolymers of different DVB contents (Jones 1952; Lloy & Alfrey 1962).

It must be noted that the selectivity of a strong base resin is closely related to the amount of water surrounding the quaternary groups (Chu *et al.* 1962). Consequently, a strong base resin with a high homogeneous water content has better kinetics and is easier to regenerate than a resin of a lower water content, having about the same wet volume capacity.

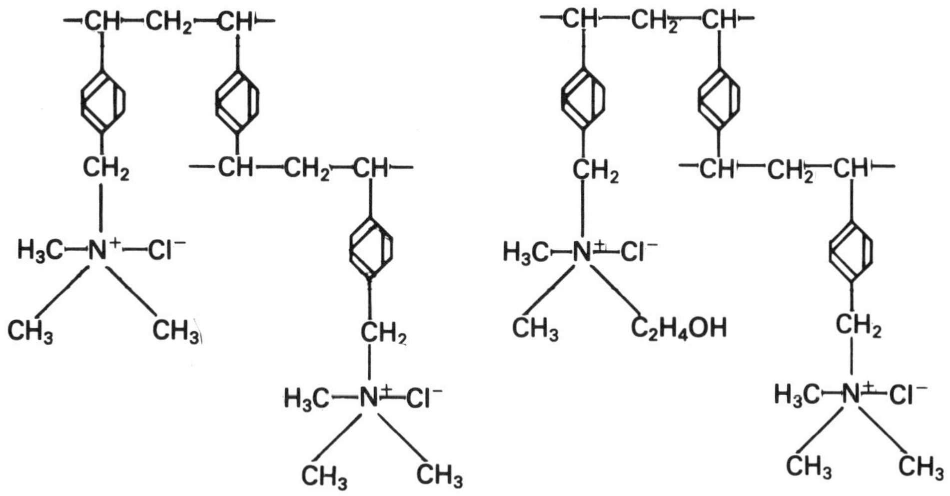

Fig. 7.5 – Strong base type I anion exchange resin (left).
Strong base type II anion exchange resin (right).

7.4.2 IER porosity

We have seen that gel resins can have various microporosities and that the size of the micropores increases with the decrease in DVB content. Macroreticular (MR) resins have a gel phase with islands of discontinuity in their gel network (Fig. 7.6).

A model structure of an ion exchange resin is given in Fig. 7.7. Fig. 7.8 illustrates the microscopic difference between gel and MR resins. Figs. 7.9 and 7.10 show an impressive close-up of a macroreticular strong base ion exchange resin (Amberlite IRA-938).

If a gel resin is dried, its internal structure can collapse and re-wetting the resin often does not completely recover the original geometry and kinetics. MR resins, owing to their special structure, can often be re-wetted with no loss of properties (Fig. 7.11).

Macroreticular resins have faster exchange kinetics and they can be operated in organic solvents (aprotic, protic, apolar and polar). In fact, the void volume inside the MR structure permits the solvent and the solute to diffuse inside the resin bead, regardless of the polarity and nature of the solvent, and therefore, ion exchange can take place, at least on the active surface of the resin circumventing the void volume. The degree of solvent penetration in the gel phase of the MR structure is influenced by the nature and polarity of the solvent and eventually also of the solute. The capacity of IERs in different solvents vary with the solvent (Fig. 7.12).

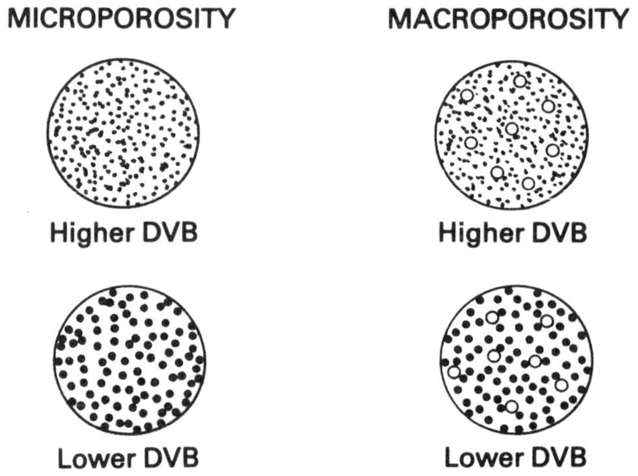

Fig. 7.6.

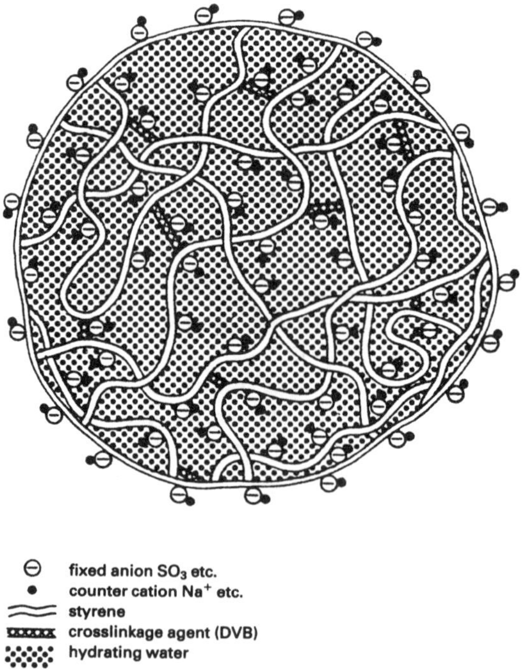

Fig. 7.7 – Model structure of ion exchange resin (Wintero, 1947).

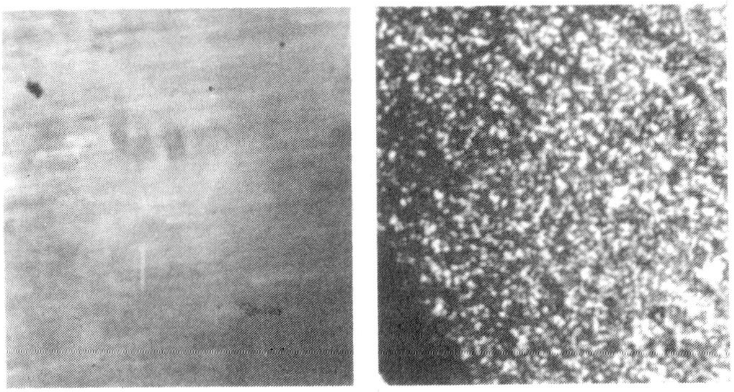

Fig. 7.8 – Electron micrographs of gel and macroreticular structures. Magnification: 600×.

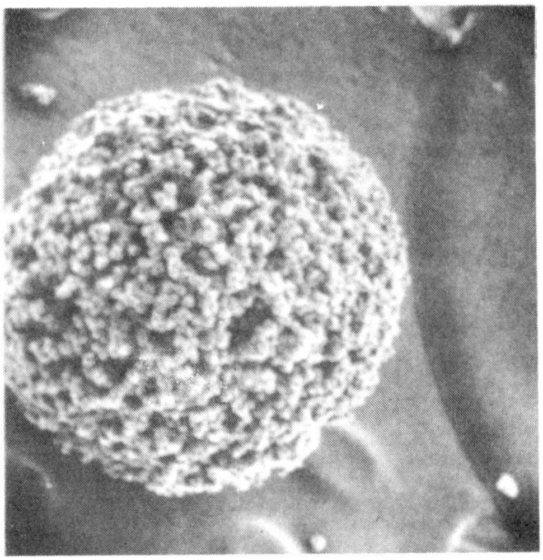

Fig. 7.9 – Amberlite IRA-938. Magnification: 500×.

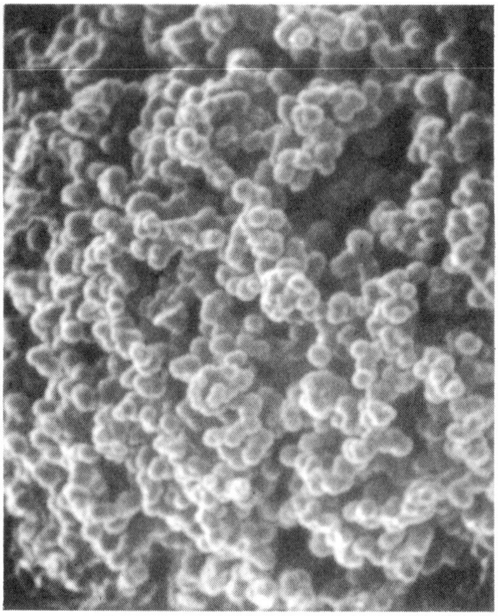

Fig. 7.10 – Amberlite IRA-938. Magnification: 960×.

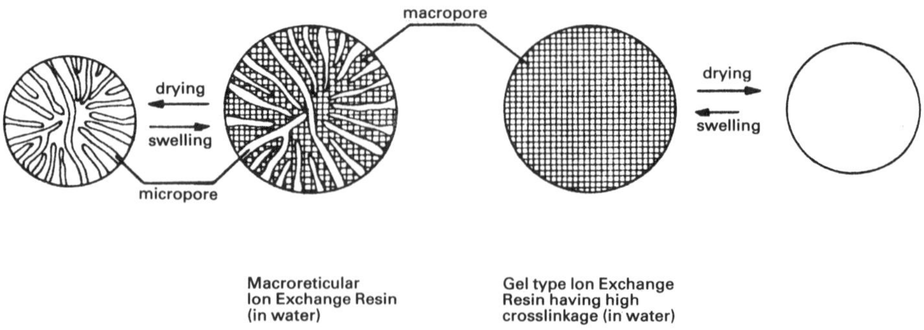

Fig. 7.11 – Effect of drying on gel and macroreticular resins.

Exchange capacity of strong acid cation exchange resin† **in solvents** (Kunin 1958).

Meq. adsorbed	Triethyl amine	Pyridine	Strontium	Ca	Pb	Co	Mn	Fe
Solvent	meq/g-R	meq/g-R	meq/g-R	meq/g-R	meq/g-R	meq/g-R	meq/g-R	meq/g-R
Water	4.21	3.93						
Ethanol	3.39	3.97	1.61					
Acetone	3.52	3.62	0.53					
Dioxane	3.80	3.80	1.26					
Benzene	3.80	3.87	1.49	3.68	4.00	1.78		
Gulfoil 361	2.80	2.20	0.39	2.64	2.92	0.31	0.57	0.19

† Amberlite IR120

Exchange capacity of strong base anion exchange resin† **in solvents** (Kunin 1958).

Meq. adsorbed	Acetic acid	Stearic acid	Octadecanoic‡ acid
Solvent	meq/g-R	meq/g-R	meq/g-R
Water	2.70		
Ethanol	2.60		1.13
Acetone	3.01		
Benzene	5.30§	0.84	0.27
Dioxane	2.12	0.46	
Gulfoil 361	6.07§		0.83¶
Total exchange capacity	2.8		

† Amberlite IRA 400 except where indicated.
‡ A highly branched isomer.
§ Indicates probable physical adsorption.
¶ Amberlite XE 75 (now Amberlite IRA 401).

Fig. 7.12.

7.4.3 Molecular exclusion

Let us imagine that the gel structure of an ion exchange resin behaves like a membrane, and that the MR structure is like a perforated membrane (Fig. 7.6), and let us extend this concept to a three dimensional network. Having this in mind, it is easy to predict that when a gel IER is put in contact with a solution of molecules having different sizes, molecular exclusion can take place (Fig. 7.13).

Different ionic forms of the same IER can give different molecular exclusion effects (resin swelling/shrinking). The polarity of the molecules in solution can affect molecular exclusion since absorption on the outer and/or the inner surface of the beads can take place (electrostatic forces). Precipitation can occur inside the beads, owing to the concentration gradient throughout the bead (Fig. 7.1).

Molecular weight is important, but even more so is the three-dimensional structure of the molecule in solution.

There are several known examples of industrial molecular exclusion used for difficult separations, i.e. streptomycin (US Patent 2,945,850) and dimethyl aniline from Penglobe† (Swedish Patent 7810122).

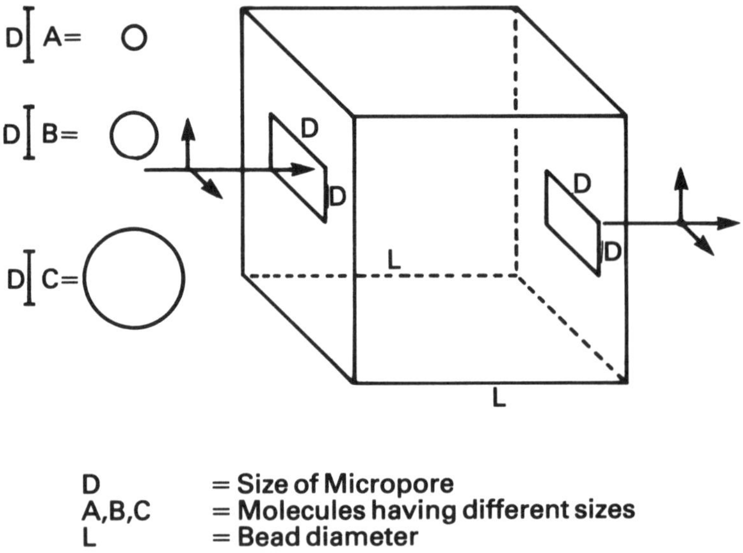

Fig. 7.13 – Molecular exclusion.

7.4.4 Ion exclusion

When an IER is put in contact with an aqueous solution of electrolytes, in general the concentration of the electrolytes inside the resin particles is lower than the concentration of the same electrolytes outside the resin. This phenomenon can be explained by Donnan membrane equilibrium (Fig. 7.14).

† Bacampicillin hydrochloride.

When compared with the result of electrolytes, the difference in concentration of non-electrolytes inside and outside the resin particles is found to be smaller (Fig. 7.14). If a mixture of an electrolyte and a non-electrolyte is slowly passed through an ion exchange resin column, the electrolyte of lower *Kd* value comes out first, followed by the non-electrolyte of higher *Kd* value. Since the coefficient of distribution for electrolytes and non-electrolytes is different, the two can be separated. For solutions of electrolytes, the *Kd* value depends upon the type and concentration of the electrolytes, but it is always lower than 1 and, in general, about 0.1.

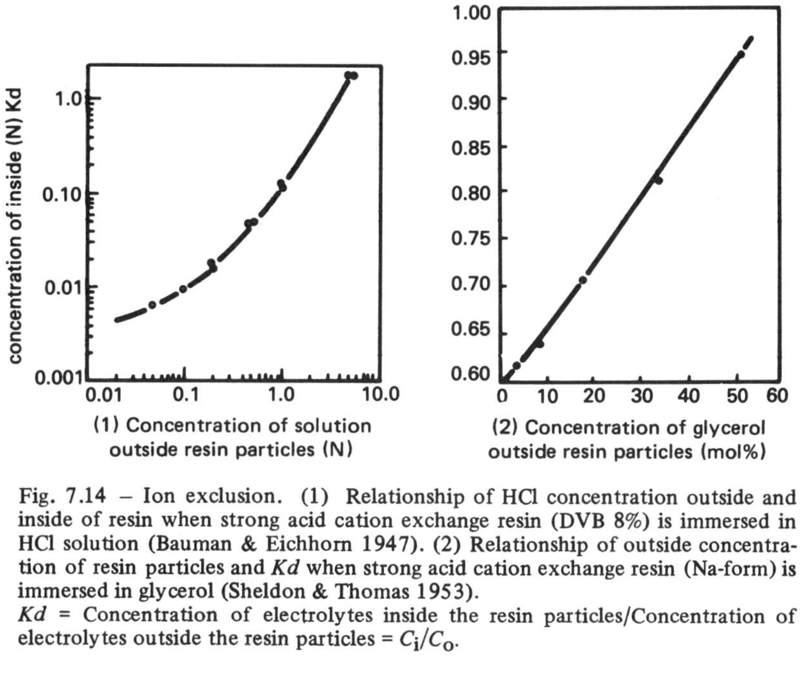

Fig. 7.14 – Ion exclusion. (1) Relationship of HCl concentration outside and inside of resin when strong acid cation exchange resin (DVB 8%) is immersed in HCl solution (Bauman & Eichhorn 1947). (2) Relationship of outside concentration of resin particles and *Kd* when strong acid cation exchange resin (Na-form) is immersed in glycerol (Sheldon & Thomas 1953).
Kd = Concentration of electrolytes inside the resin particles/Concentration of electrolytes outside the resin particles = C_i/C_o.

7.5 FOULING OF THE MATRIX

There are several products commonly found in filtered antibiotic fermentation broths which can foul ion exchange resin matrices: anti-foaming agents, proteins, waxes, endogenous Ca, suspended solids, fatty acids, oils, phenolic substances, thiols, etc.

The concepts reported here apply to gel and MR ion exchange matrices and also to *polymeric adsorbents* which will be discussed later.

7.5.1 Anti-foaming agents, waxes, oils

Most of the substances are hydrophobic and, consequently, can be adsorbed on hydrophobic matrices through electrostatic attraction. This phenomenon can be particularly important in the case of styrene–DVB matrices.

Below are listed the most important factors which influence the electrostatic adsorption of the above-mentioned products on matrices and adsorbents and some pragmatic suggestions for regeneration of the polymer: nature and molecular size of the anti-foam, wax, oil; resin porosity; contact time with the resin; temperature; ionic strength of the solution (salting-out).

If this type of adsorption has been encountered, then to regenerate the resin, the following practical alternatives can be considered: elute the adsorbed product with a suitable organic solvent; use a dilute regenerant solution at the highest possible temperature; hydrolyse the adsorbed product by means of chemical or biological (enzymic) agents; use a suitable surfactant.

7.5.2 Suspended solids

When a filtered broth is passed over an ion exchange resin or adsorbent and the wanted product is exchanged with, or adsorbed by, the resin, the composition of the broth changes and often precipitation occurs.

In such cases it is often difficult to economically prevent such phenomena; sometimes ultrafiltration does the job, but it can be expensive.

The most simple solution is, of course, backwashing, but if the precipitation rate is somewhat slow, simple backwashing might not be sufficient. In this case, it is worth considering backwashing the resin with a suitable low foaming surfactant to reduce possible electrostatic interaction between the solids and the resin matrix. Backwashing with a 100% bed expansion is normally worth considering.

In some cases, the solids dissolve by simply changing the pH. Solids must be removed as frequently as economically possible in order to prevent more serious problems such as, for example, microbial infection in the resin column, resin clumping, bed channelling, etc.

7.5.3 Fatty acids, phenolic substances, thiols

These substances can react with anionic groups of ion exchange resins and/or be adsorbed on the surfaces by electrostatic forces. If they are not eluted by standard regeneration procedures, one should consider using a less concentrated regenerant such as, for example, 2% NaOH, or even 0.5% NaOH, instead of the standard 4% NaOH, in order to reduce the salting-out phenomenon. Another alternative is to use an aqueous solvent solution of regenerants, or even pure organic solvent. Wrong regeneration conditions could precipitate substances such as, for example, polyphenols and irreversibly foul the resin.

Phenolic substances and, even more, thiols, can polymerise and film-coat fouled resin. In this case, oxidising treatment of a regenerated resin should be considered in order to break the formed film, followed by at least another regeneration prior to reusing the resin.

Oxidising treatments, such as, NaClO, H_2O_2, etc. should have the aim of reducing the molecular weight of substances that could not otherwise be eluted. By no means should an oxidising treatment be used as a type of regeneration. In the case of NaClO, a concentration of 0.5–5.0 ppm, active Cl_2 is normally used. Treatment is considered complete when the concentration of Cl_2 in the outlet is 80–90% of the concentration at the inlet.

In general, the oxidisable organics protect resin matrices from oxidation and often even protect the column construction material from oxidation. There are known cases where 316 stainless steel columns, containing fouled resin, have successfully been treated a few times a year for over six years, without any indication of column corrosion or resin destruction.

The presence of heavy metals, such as Fe, for example, must be avoided; these heavy metals could act as catalysts for radical de-crosslinking reactions.

7.5.4 Proteins

These are perhaps the most difficult and certainly the most unpredictable substances. Proteins react with all types of ion exchange resins and adsorbents, but to different extents and with different mechanisms. Proteins also tend to react with almost all types of the previously described impurities present in the fermentation liquors. The treatments for the elution of protein from the resin are all those described in sections 7.5.1–7.5.3, plus combinations of them.

A technique which is still under evaluation consists in the formation of soluble complexes between proteins and glycosides and/or heavy metals.

7.6 IMPORTANCE OF PARTICLE SIZE AND PARTICLE SIZE DISTRIBUTION IN THE SEPARATION OF ANTIBIOTICS

When dealing with a fermentation broth, the major concern is that the resin extracts the antibiotic which can then be eluted easily, and that the resin is regenerable. Standard resins meet these needs. However, sometimes, people look for more sophisticated separations where the kinetics of standard resins are too slow under normal operating conditions.

One example of such a need is the chromatographic separation of gentamicin (Pirotta 1980) where special resins have to be used. A combination of small radii and very good particle uniformity causes each chromatographic resin particle to react with similar kinetic values. By means of these special grades, it is often possible to obtain separations that would be impossible with standard particle size resins (Pirotta 1980; Box *et al.* 1979; Hood *et al.* 1979).

7.7 ADSORBENTS (Amberlite XAD family)

Adsorbents are MR matrices with very high active surface and porosity. Although all the previously mentioned points applicable to matrices are also applicable to adsorbents, these products are not exactly pure matrices and, therefore, we here illustrate some specific properties of Amberlite XAD adsorbents. Although the most commonly used adsorbents are based on styrene–DVB copolymers or acrylates, we shall refer here only to the styrene–DVB and poly–DVB adsorbents.

The pore size of an adsorbent plays an important role (molecular exclusion and adsorption stability) in the extraction of antibiotics from fermentation broths. In an adsorbent the definition of its pore size and its pore size distribution is statistically determined (He penetration – Hg penetration method); consequently, the values so obtained cannot indicate where the 'porosity' is located. If, paradoxically, all the small pores were on the outer surface of the bead and the large ones in the inner active surface, it would be quite possible that molecular exclusion would prevent any significant adsorption of antibiotics from taking place (Pirotta 1982). Nevertheless, the pore size distribution and the average pore size curves would indicate that the adsorption is highly possible. The following table illustrates that the definition of porosity (Fig. 7.15) is normally considered to be two-dimensional.

	Width (in Å)
Micropores	Less than 20
Mesopores†	Between 20 and 500
Macropores	Above 500

† This replaces the earlier terms 'intermediate pores' and 'transitional pores'.

Fig. 7.15 – Classification of pores according to their width (Albright, forthcoming).

If we wish to predict whether an adsorbent can be used successfully to extract an antibiotic from its fermentation broth, the 'porosity' factor of the adsorbent has to be seen as three-dimensional. Consequently, pore shapes and pore volumes are important factors in separations. It is important to emphasise that the water and, therefore, the solute to be adsorbed, is located in the pore volumes of the adsorbent and not coordinated to the copolymer – therefore, speaking of moisture-holding capacity for an adsorbent is semantically wrong. However, this imprecision is normally accepted as a similitude of IER jargon.

The active surface involved in the adsorption is the surface in contact with the water-solute system only, and is a function of the percentage of aromatic rings (poly–DVB adsorbents) (Fig. 7.16) and of the microsphere diameter (Fig. 7.17).

Starting from the same monomers, the polarity of the active surface depends on the copolymerisation system used. Particularly, polarity depends on the number of unsaturated aliphatic chain terminals. The polarity of a poly–DVB adsorbent can be changed by other means such as, the addition of groups such as Br, Cl, NO_2, etc., either to the copolymer or to the monomers, prior to copolymerisation, or by changing monomer composition, etc.

For porous aromatic polymers, the surface capacity only becomes significant when approximately 25% of the aromatic rings are on the internal surface of the particle.

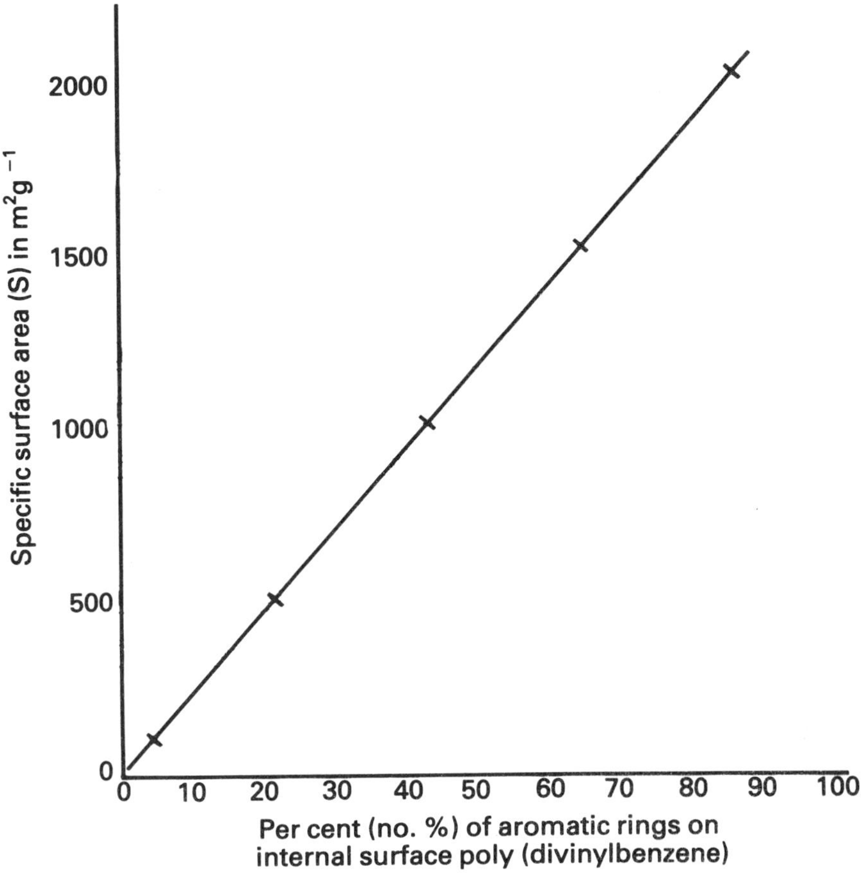

Fig. 7.16 – Number per cent of aromatic rings on internal surface of porous poly (divinylbenzene) as a function of specific surface area (Albright, forthcoming).

The specific surface for a porous aromatic polymer with 25% of its aromatic rings on the terminal surface is about $500\,m^2\,cm^{-3}$ ($540\,m^2\,g^{-1}$ at a skeletal density of $1.080\,g\,cm^{-3}$) (Albright, forthcoming).

Molecular size, shape and intrinsic nature of the product to be adsorbed very markedly influence the effectiveness of a specific porous polymer as far as its adsorption chracteristics are concerned (Albright, forthcoming).

Several antibiotics have been industrially extracted by means of Amberlite XAD adsorbents for many years, for example, cephalosporin C (US Patent 3,725,400) tetracycline, etc.

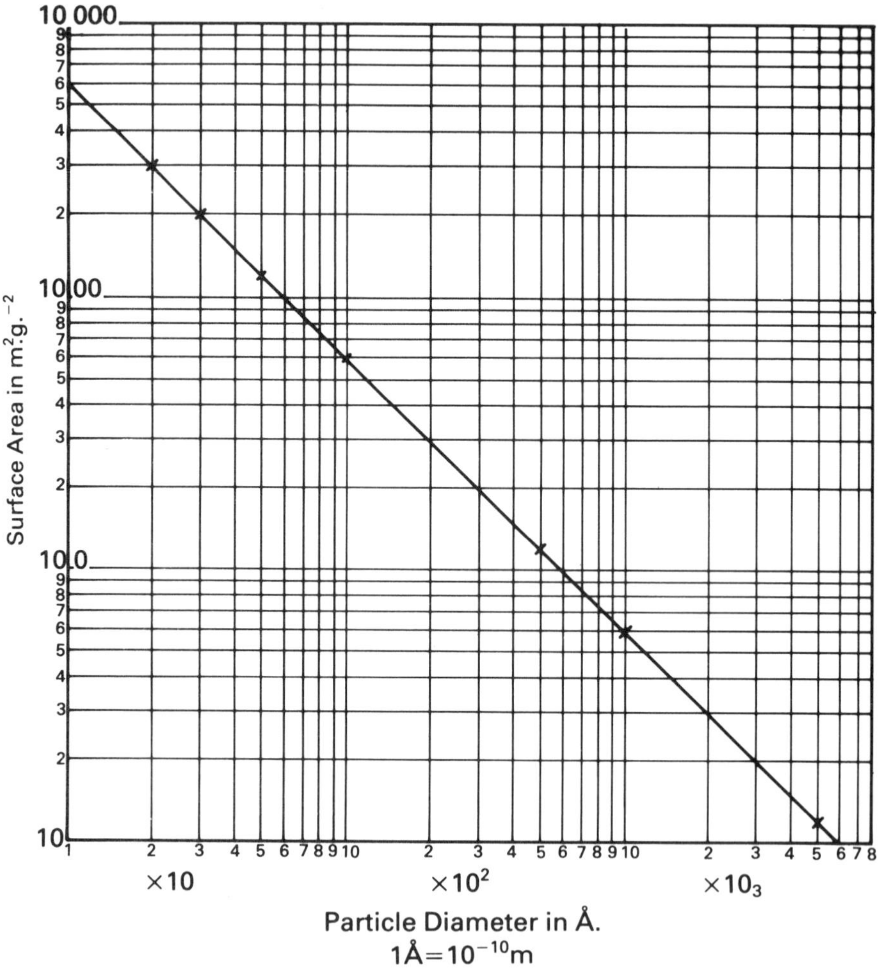

Fig. 7.17 – Surface area as a function of microsphere diameter (Albright, forthcoming).

REFERENCES

Albright, R. L. (forthcoming) Porous polymers as an anchor for catalytic agents to perform chemical work.

Baumann, W. C. & Eichorn, J. (1947) *J. Am. Chem. Soc.* **69**, 2830.

Box, S. J., Hood, J. D. & Spear, S. R. (1979) *J. Antibiotics* **32**, No. 12, 1239–1247.

Chu, B., Whitney, D. C. & Diamond, R. M. (1962) *J. Inorganic Chem.* **24**, 1405.

Hood, J. D., Box, S. J. & Verrall, M. S. (1979) *J. Antibiotics* **32**, No. 4, 295–304.

Jones, G. D. (1952) *Ind. Eng. Chem.* 44, 2686.

Kunin, R. (1958) *Ion Exchange Resins.* Chichester, John Wiley. pp. 310–315.

Lloy, W. G. & Alfrey, J. (1962) *Pol. Sci.* **62**, 301.

Pirotta, M. (1980) Abstracts 4th Symposium on Ion Exchange Resins, Balaton Lake, Hungary, 27–30 May. Hungarian Chemical Society, pp. 241–242.

Pirotta, M. (1982) *Die. Ang. Makr. Chem.* **109/110**, No. 1774, 197–214.

Sasaki, I., Gotoh, H., Yamamoto, R., Hasegawa, H., Yamashita, J. & Horio, T. (1979) *J. Biochem (Tokyo)* **86**, No. 5, 1537–1548.

Sheldon, J. & Thomas, L. H. (1953) *J. Appl. Phy.* **24**. 235.

Swedish Patent SE-7810122 (Astra Sodertaljever).

US Patent 2,945,850 (Olin Mathieson Chemical Corporation).

US Patent 3,725,400 (Ciba-Geigy Corporation).

Wintero, J. C. (1947), *Chem. Ind.* **62**, 745.

8

Scale-up of isolation processes, approaches and problems

Walter Voser and Hans-Paul Walliser
CIBA-GEIGY AG, CH-4002 Basel, Switzerland

8.1 INTRODUCTION

'Biotechnology scale-up: How big a problem?'. This question posed in *Chemistry Engineering News* (Baum 1984) precedes the following passage: 'It has become an article of conventional wisdom that scale-up to commercial production is one of the major problems facing the biotechnology industry. The industry's products, almost exclusively chemically sensitive macromolecules, must be separated from the cells that produce them and taken to a high level of purity. That's not too difficult at the lab bench, but many observers have speculated that it would be a stumbling block on a commercial scale'.

Two representatives of firms practising the 'new' biotechnology dispute that notion. To quote their views on this statement, they say: 'That's folklore. Somewhere along the line that idea has crept in, but it is unknown where it came from. The problems that do arise will be of the type that any new production process experiences and will yield to the same types of solutions. The technology is not years away. It exists'.

These conflicting points of view for scale-up of recovery processes for products from recombinant organisms can be heard whenever this problem is discussed. It is very likely that similar statements were made at the beginning of the era of modern biotechnology, when chemists were confronted with the recovery of penicillin, a relatively unstable compound, streptomycin and other antibiotics from fermentation broth. In the meantime these technical problems have been solved and this type of industrial production has been functioning for decades. Yet scale-up of processes for isolation of new microbial products is a real challenge for chemists and chemical engineers. Such processes can be quite simple or very complex, depending on the intended use of the products, be it as a pharmaceutical, an intermediate or a feed-additive.

8.2 NATURE OF THE PROBLEM

Some forty years of experience in the recovery of microbial products on an industrial scale has led to the accumulation of considerable know-how. Nevertheless, in spite of all the successes achieved in scale-up, the early development of a recovery procedure for a new substance can be difficult, particularly in the first step from the laboratory to pilot level. There are a number of reasons for these initial difficulties:

- titres at the beginning of process development are normally rather low;
- fermentation conditions are not yet optimised;
- the nutrient medium tends to be more complex than after optimisation;
- in general, the development of fermentation and recovery procedures occurs in parallel with changes in nutrient medium composition and the introduction of new strains;
- batch to batch variation, often caused by microbial contamination, is unfortunately the order of the day;
- scaling-up can be a scaling-down problem – that is to say, the relevant pilot-scale equipment may not be available or is much less efficient or difficult to handle than that which is designed for industrial production, and not all industrial parameters can be simulated on pilot scale; sometimes insufficient product or culture broth volume is available, especially in a multistep procedure when starting with low titre fermentation broth;
- development and scale-up of a recovery process for an existing industrial plant may necessitate compromises in process strategy, and
- development of the fermentation process never ceases.

Positive aspects are:

- in general, the efficiency of recovery processes increases with increasing titres, and
- by close cooperation between the recovery specialists and the microbiologist, reciprocal process improvement can be achieved.

For the recovery of microbial products all the separation methods applicable for isolation and purification of organic substances can be used. The most interesting ones from the point of view of scale-up are: solid–liquid separation, extraction, chromatography and concentration.

The numerous unit operations, separation methods and the available equipment to perform them are the tools for scaling-up. The choice of process stategy is paramount when developing an optimal procedure. In addition, a careful evaluation of all process parameters and a sufficiently long piloting time must be allowed to be sure that the final industrial process is flexible enough to deal with natural fermentation batch variations.

It is impossible to cover even briefly the relevant scale-up aspects of the four separation methods already referred to. An excellent review with numerous references was published by H. Gerstenberg *et al.* (1980). We intend to confine ourselves to some of the scaling-up aspects of chromatography including column design.

8.3 CHROMATOGRAPHIC PROCEDURES

In general, problems encountered in the scale-up of solvent extraction procedures for the isolation of microbial products can be said to be relatively simple compared with scale-up of chromatographic procedures which can still be cumbersome. This is not surprising because chromatographic procedures are generally much more complex and not all parameters can be studied at once on pilot scale. It is obviously impossible to simulate the influence of bed height and bed diameter on the separation efficiency and physical phenomena. Also some problems on pilot scale are more serious or different than on industrial scale, in other words scale-down problems can occur here.

The wide variety of chromatographic methods ((Gerstenberg *et al.* 1980; Voser 1982); Table 8.1) and of physicochemical properties of suitable adsorbents for industrial-scale chromatography is responsible for their broad spectrum of adsorption and hydraulic properties. The adsorbents can be natural or synthetic, inorganic or organic, neutral or charged, highly polar to non-polar, soft to rigid, gel-type to macroporous, fine to coarse, fibrous to grained, granular or globular, swelling to non-swelling.

Table 8.1 – Chromatographic methods (incomplete).

Chromatography method	Operating principle	Separation by differences in
Adsorption	Surface binding by van der Waal's forces	Polarity
Affinity	Biospecific complex formation	Macromolecular structure
Covalent	Weak covalent forces	Reactive SH-groups
Gel	Pore diffusion	Molecular size
Ion exchange	Ion binding	Charge
Distribution	Distribution coefficient between two liquid phases	Polarity, solubility
Ligand exchange	Coordination complexes	Complex formation with transition metals

The aim of scaling-up a chromatographic processes is to obtain the same yield and product quality in the same period of time on laboratory, pilot and industrial scale. Volume scale-up in chromatographic processes is straightforward. The first step in scaling up is the choice of the process strategy. It can be downflow, upflow or mixed, successive cycles with or without intermediate backwashing, backwashing at longer intervals or no backwashing at all, unloading for special regeneration, operation with positive head pressure or hydrostatic pressure only, single columns or columns in series or batch or continuous operation. The next step is the evaluation of the maximal and optimal bed height. For the determination of the maximum feasible bed height, the optimal specific volume velocity, expressed in bed volumes per hour and evaluated in laboratory experiments, is maintained constant. The limiting factors will be either pressure drop during the most critical phase or unfavourable adsorption/desorption kinetics, since, with increasing bed height, the linear velocity increases too. Bed heights varying from 15 cm to 12 m have been reported. Once volume velocity and bed height have been fixed, column scale-up is straightforward and the suitable column diameter is selected to give the desired bed volume. Sample volume or loading must be kept constant for maximum efficiency depending on the mechanism of separation.

Process strategy and bed height determine the column design. Whether the resulting column dimensions are feasible or smaller columns in parallel must be operated is determined by practical considerations such as construction problems, available space and flexibility. One crucial requirement for any column system is the uniform distribution of flow from the inlet and outlet tubes across the entire width of the column bed combined with minimal mixing and dilution in the flow distribution chambers at either end of the column.

Scale-up problems differ according to whether chromatography is used as the first step for initial purification, or later on in the process for final purification. As a rule, in the first case huge volumes and extremely complex solutions are involved. This necessitates the use of high flow rates and efficient regeneration procedures. In the second case high separation efficiencies are required and regeneration is no problem or only a minor one. To meet these requirements very precise column design and construction is needed.

8.4 APPLICATIONS

8.4.1 Low molecular weight products

Good examples of processes in which enrichment of microbial products by chromatographic methods on industrial scale are used are the recovery of streptomycin and cephalosporin C. Both are hydrophilic substances but streptomycin is a strong base whereas cephalosporin C is a weak acid. The strong and weak ionic nature of the molecules determines the strategy for their recovery. Weakly acidic, acrylic-type cation exchangers have a high capacity for streptomycin (Hoffman *et al.* 1957; US Patent 2,528,188). High loading from quite impure solutions can be achieved, because most of the cationic by-products are displaced during adsorption because of the strongly basic nature of streptomycin. These favourable adsorption characteristics yield eluates of considerable purity. The favourable physicochemical properties of streptomycin are ideal for scaling-up. In principle, all the sophisticated batch or continuous ion exchange methods developed for water treatment can be used, if an elution step is introduced.

One of the most satisfactory achievements for the chemical engineer was the successful recovery of streptomycin from the unfiltered broth on Amberlite IRC-50 by upflow adsorption. A number of measures are mentioned in the literature to avoid flooding caused by the small density difference between broth and resin. Such measures are dilution of culture broth, when necessary, low linear flow rates, reduction of the linear flow rate in the upper part of the column by increasing the column diameter or the installation of a screen and a slowly rotating propeller in front of the exit pipe to prevent the resin from escaping the column (Bartels *et al.* 1958). In any case, two to three columns in series are essential to obtain high yields.

Weakly basic acrylic type anion exchangers also have a reasonably high capacity for cephalosporin C. However, in contrast to streptomycin, the weakly acidic cephalosporin C is displaced by stronger acidic impurities during percolation of culture broth. The resulting capacity can then drop to 10–20% of nominal. The means that an efficient enrichment step is unobtainable. The unfavourable properties preclude the use of the whole broth technique.

For enrichment of charged microbial products in general and cephalosporin C in particular, uncharged macroporous styrene–divinylbenzene type adsorbents of the Amberlite XAD-type can be used very efficiently as an alternative to ion exchangers (Voser 1982; US Patent 3,522,248). However, for scaling-up there is a negative aspect, the capacity of the resins is relatively low. For pure cephalosporin C it is 20 to 100 g dm^{-3} depending on the resin, the concentration of the antibiotic and other factors. For cephalosporin C in culture filtrates the maximum possible loading is lower and depends on the production strain, the general fermentation conditions and the titre. As the highly polar cephalosporin C is adsorbed by van der Waal's forces the actual capacity of the macroporous adsorbents depends among other things on the amount of less polar impurities present which have higher affinity to the resin. Positive aspects which favour the use of macroporous adsorbents for recovery of cephalosporin C and other microbial products are their adsorption/desorption kinetics as well as their high chemical and physical stability. An adsorption/elution/regeneration cycle tends to be relatively short, i.e. 3–6 hours, and hundreds of cycles can be performed with minimal loss of resin. All these facts dictate and allow the following process strategy for industrial production. Each batch is processed in several portions to reduce the necessary resin volume. Downflow operation is chosen throughout, including regeneration, because backwashing is time-consuming. Of course, backwashing will be necessary but only at longer intervals. Adequate automation and extensive pilot studies are advisable. Special attention has to be paid to the optimisation of the resin regeneration and to the maintenance of a homogeneous resin bed from one cycle to the next. Problems with homogeneity can be connected with the resin. Therefore careful evaluation of a suitable resin is essential for scaling-up, whereby not only capacity and adsorption/elution kinetics have to be taken into consideration but also physical parameters such as swelling and radial pressure on the column wall. Extensive swelling can be problematic for column design and high radial pressure can be troublesome on pilot scale for resin bed homogeneity and dangerous when glass columns are involved. One might expect that swelling and radial pressure are strictly correlated, but this is not the case, as the following evaluation of resins shows. The goal of the evaluation was to select a resin for industrial use with about twice the capacity of Amberlite XAD-2.

The resins were selected on the basis of their specific inner surface. Some were eliminated for reasons of capacity and one of them because unavoidable channelling of the resin bed occurred under actual working conditions during displacement of the regenerant solution. The other resins gave the rather surprising results shown in comparison with Amberlite XAD-2 in Table 8.2. The radial pressure was measured three column diameters below the resin surface. The values in brackets are from the manufacturer, the others were determined by Ciba-Geigy. Swelling is measured after backwashing.

Table 8.2 – Resin parameters – radial pressure and pressure drop.

	Amberlite XAD-2	Amberlite XAD-4	Diaion HP-20	Amberlite XAD-1180
1. Spec. surface ($m^2 g^{-1}$)	326 (300)	702 (781)	560 (718)	640
2. Spec. pore volume ($cm^3 g^{-1}$)	0.48 (0.98)	0.65 (0.98)	0.99 (1.16)	1.62
3. Rel. swelling	1	1.7	2.5	1.5
4. Radial pressure Max. values				
1. Cycle (bar)	3.3	5.0	3.2	1.0
2. Cycle (bar)	3.3	5.2	4.4	1.0
3. Cycle (bar)	3.3	5.4	5.4	1.0
4. Cycle (bar)	3.3	5.6	6.2	1.0
5. Cycle (bar)	3.3	5.8	7.0	1.0
5. Pressure drop Max. values (bar)	0.1	0.2	0.3–0.7	0.1

Bed dimensions: $d = 245$ mm; $h = 860$ mm.
Flow rates: 2 BV h^{-1} (XAD-2 1.5 BV h^{-1}).

– Amberlite XAD-2 showed quite a high radial pressure of 3.3 bar when changing from water to water containing 45% propan-2-ol and these measurements remained constant from cycle to cycle. The maximum and constant pressure drop was 0.1 bar. The swelling was 9% and in Table 8.2 is expressed as 1.

- Amberlite XAD-4, with a 1.7 times greater swelling, developed a considerable radial pressure of 5 bar during the first cycle which increased slightly by 0.2 bar from cycle to cycle. The pressure drop was 0.2 bar throughout.
- Diaion HP-20 showed a 2.5 times higher swelling ratio than Amberlite XAD-2 but with 3.2 bar the same radial pressure as Amberlite XAD-2 in the first cycle which increased steadily from cycle to cycle. After five cycles the radial pressure was 7 bar and still increasing. The pressure drop increased in parallel from cycle to cycle, starting at 0.3 bar and after five cycles was 0.7 bar.
- Amberlite XAD-1180, which at the time of use was an experimental resin and is identical with Amberlite ER-180 described in the literature (Pirotta 1982), had a moderate swelling factor of 1.5, a very low radial pressure of 1.0 bar which was constant over five cycles, and a low pressure drop of of 0.1 bar.

The result of this resin evaluation was astonishing. The selected resin Amberlite XAD-1180 was ideal for the chosen process strategy on pilot scale to perform one cycle after another without backwashing. Its favourable properties greatly facilitated pilot scale work. Increase of radial pressure on the small scale may be a problem but its effect on an industrial scale is not yet known, whereas in the latter case swelling may be the major problem.

In the older literature two main systems for industrial columns suitable for use with beaded synthetic organic ion exchangers and macroporous adsorbents were described. One is based on subdivision of the column into sectors with baffles or other systems to avoid ‘fingers’, etc. (*Chem. Eng. News* 1968; US Patents 3,250,058, 3,539,503). The other, used for sugar separations, envisages floating distributors to achieve an even distribution of the feed solution and a minimum volume of supernatant (US Patent 3,814,253; German Patent 2,306,210). The bottom construction is conventional, but with a low dead volume. However, with subdivided columns, backwashing, which is particularly necessary when the chromatographic step is used for enrichment of microbial products, is impossible and, in the case of floating distributors, tends to be complicated.

Two other construction principles might be more suitable for use in initial purification steps. Both are operated with positive head pressure by which, in addition to other advantages, gassing out inside the column is avoided.

Relatively slender columns (Fig. 8.1) with height/diameter ratios higher than 2:1 are ideal for pilot work and industrial scale when the bed height is not a limiting factor and the liquid level is kept constant. At maximum swelling, the resin should remain just covered with liquid. For level control a laterally mounted differential pressure transmitter which acts on the bottom valve of the column is recommended. The feed system is a straight pipe distributor with adequate perforation to allow mixing in the supernatant while changing from one phase to another without disturbing the resin surface. The bottom construction is

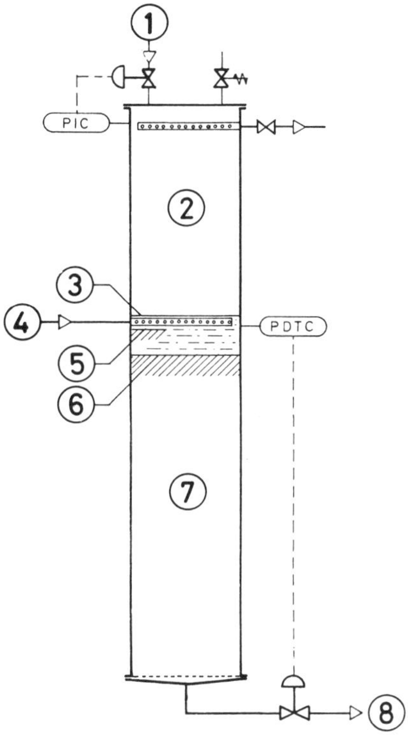

Fig. 8.1 – Construction principle of relatively slender columns (liquid level constant). (1) Nitrogen; (2) Empty space for backwashing; (3) Liquid level; (4) Feed line and sparger; (5) Upper resin level; (6) Lower resin level; (7) Resin bed; (8) Exit line.

conventional, with a low dead volume. For some applications this construction has the disadvantage that relatively high mixing volumes must be accepted at the lowest resin level and at low height/diameter ratios, unless this can be avoided by stratification using liquid density differences.

'Short squat' columns (Fig. 8.2) with a height/diameter ratio lower than 2:1 are best adapted for industrial use. Rohm & Haas Europe has developed a favourable concept and the relevant know-how for an arrangement which is especially suited to low volume velocities. The top and bottom construction is based on hydraulic considerations and guarantees an even distribution of the feed solution together with collection of the percolate without disturbing the flow patterns within the bed. The liquid level is controlled by a differential pressure transmitter in such a way that while changing from one phase to another, the supernatant is minimal. The pressure transmitter is mounted in the column wall below the resin surface and acts on the exit valve. This construction should guarantee horizontal zones inside the resin bed and therefore an optimal separation efficiency. Resin bed dimensions of 3–4 m diameter and 1–2 m height should not cause problems, even for relatively fine resins. However, it is not yet known whether columns of this type require modifications when used for many cycles without backwashing of the resin.

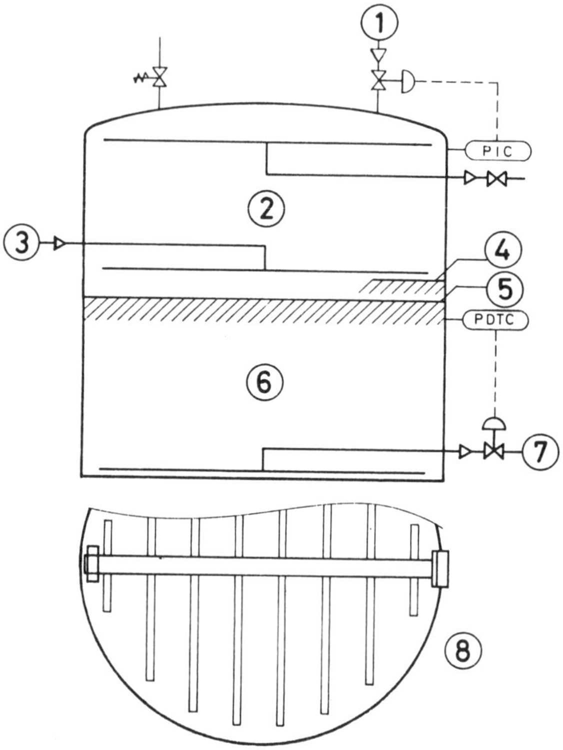

Fig. 8.2 – Construction principle of 'short, squat' columns (Rohm & Haas) (liquid level adjustable by programming). (1) Nitrogen; (2) Empty space for backwashing; (3) Feed line; (4) Upper resin level; (5) Lower resin level; (6) Resin bed; (7) Exit line; (8) Top and bottom distributor.

8.4.2 Macromolecular products

An important fact concerning pilot and industrial scale chromatographic columns is that no column construction is broadly applicable or ideal for large numbers of different types of adsorbent. The universal large-scale column simply does not exist. This becomes especially evident when fine or soft adsorbents are involved, especially for final purification of enriched products. This is more or less the rule in the recovery of proteins and other high molecular weight products. The maximum bed height at the optimal volume velocity is then rather low, often only 15–20 cm, and scaling-up is achieved by increasing the column diameter and using stack columns. Achieving even distribution of the feed solution over the entire resin bed surface is also one of the main problems. The problem is even greater when gradient elution and high resolution are necessary.

Interesting approaches for handling fine and soft adsorbents on large scale have been commercialised by Pharmacia, Whatman and Amicon/Wright and an entirely unconventional solution to the problem of ion exchange chromatography has been developed by AMF Molecular Separations Division.

Pharmacia,† the first manufacturer of a suitable column for soft gels, splits the flow and distributes it through six ports on the column end plates (Ek 1968; Janson & Dunnill 1974). At the entrance of each stream an anti-jetting device spreads the liquid over a fine mesh net. A coarse net between the net and the end plate acts as support and spacer. For the soft Sephadex gels G-50 to G-200, the maximum feasible bed height is 15 cm. For scaling-up as many as six such squat columns are encountered in series. This arrangement is known as a stack column (Strobel 1982). Depending on the adsorbent filling it can be cumbersome. Also splitting of the flow can cause problems for cleaning and air entrapment, which is disastrous. For piloting a process, handling can be laborious, particularly when frequent unloading is necessary, because *in situ* regeneration for many adsorbents is not possible.

In Amicon/Wright columns the flow is distributed through a carefully designed system of radial ribs cut out of the end plates. There is a single central port with a suitable anti-jetting device to reduce and divert the high velocity of the entering stream. The adsorbent bed is covered with a sinter. Sinters are claimed to be more efficient than meshes or nets to achieve an even distribution. The larger the pore size of a sinter the less efficient the system. One drawback of sinters is their tendency to adsorb substances on their very large surfaces. In the case of crude solutions, fouling can occur, with nasty cleaning problems; and when processing relatively clean feed streams losses of activity by adsorption can be a problem. The Amicon/Wright columns are also stackable. A newly developed column with a continuously adjustable end plate is a very interesting approach but unsuitable for soft gels.

Whatman has developed a column with a new flow distribution system especially suitable for their cellulosic ion exchangers. The bed height is 18 cm. The diameter of their first commercial column is 400 mm and contains 25 l ion exchanger. The resin bed is covered with a perforated plate with a high free surface. The slightly conical head plate covers an empty space and has a steep cone in the middle. The total empty head space is about 5% of the bed volume. The feed solution enters the steep cone tangentially in its upper part. The resulting rotary movement efficiently mixes the supernatant liquid and allows gradient elution. It is claimed, that filling and equilibration only takes one hour. This principle is quite new and as yet little experience with it has been acquired. However, the approach is promising, especially as there is no screen or sintered plate in the top of the resin bed which could cause difficulties.

The Zeta-Prep cartridges are an interesting unconventional new approach for ion exchange chromatography which have been developed by AMF. The cartridges consist of concentric polymer screens, bearing the ionic groups and supported by cellulosic sheets. The flow is radial from the outer rim toward a perforated central pipe. The available nominal cartridge lengths are from 30 mm to 720 mm with a constant diameter of about 70 mm throughout. Scaling-up is therefore straightforward. Single cartridges, each mounted in a housing, can be combined to a multi-cartridge system. For such a system flow rates up to 12 l min^{-1} and BSA capacities of 1400 g are claimed. At present cartridges are

† Technical literature on all commercial columns mentioned can be obtained from the manufacturers.

available for DEAE, QAE and SP ion exchange chromatography. According to the technical bulletin this system has numerous advantageous over conventional chromatographic columns. It might be especially useful for product enrichment from crude fermentation broth but it is also claimed to be good for separation of components. It will be interesting to see how this concept is incorporated into the chromatographic separation repertoire.

8.4.3 Preparative HPLC

'Scaling-up', 'industrial scale', 'process scale separation' etc. are very vague terms with respect to volumes and product quantities. The range is from some millilitres to cubic metres and from grams to several hundred kilograms per separation cycle. For very highly active pharmaceuticals, production of kilogram quantities per batch may be the industrial scale. Therefore since high performance liquid chromatography has revolutionised analytical procedures big efforts are being made to scale-up this method. More and more promising systems are already or will be shortly on the market. This should allow preparative separations in gram to kilogram quantities. As cycle times are mostly short the columns can be quite small. All leading HPLC producers are very active in this field and offer a wide variety of systems and adsorbents. An interesting and promising approach to improve separation efficiency and to minimise the problems of column inhomogeneity during shrinking and swelling is being made by Waters Preparative and Process Division (Shih *et al.* 1983; *Chem. Proc.* 1983; Burgoyne *et al.* 1984) with their radial compressed columns. Apart from smaller units, Waters offer a 'Kilo-Prep System'. For a process engineer it will be important that such systems have a modular arrangement for ordering. Furthermore it will be essential that the columns can be unloaded *in situ* and do not have to be returned to the manufacturer for this purpose.

One general trend is that manufacturers tend to exaggerate the versatility of their columns and purification systems. Manufacturers making such general claims for their products that 'High purity capability (99%+) [sic!] eliminates the need for prior or additional time-consuming process purification steps' seem rather optimistic. Experience shows, that this can only be true in very few cases. It is hardly ever true when intracellular products have to be processed. It must be emphasised that frequent replacement of the adsorbent because of fouling is hardly feasible in industrial scale. Fouling is the big problem with soft and fine adsorbents when crude solutions for enrichment of a product are processed. The fouling is increased when for any reason the feed stream is not absolutely clear or if impurities tend to precipitate out when in contact with the adsorbent. Regeneration may then be a problem, because there is usually no empty space for backwashing and the adsorbent may shrink or swell during regeneration. Therefore, for processing of crude solutions, which may be clear, partially clarified or not clarified at all, batch adsorption may be advisable. Relatively cheap suitable adsorbents can be used for this purpose.

8.5 SUMMARY

It has not been possible to cover the entire complex field of scale-up of chromatographic procedures in the space available; but this chapter will be concluded with the following general message.

Essential to the success of recovery and purification processes is the close cooperation of the recovery specialist and microbiologist. Such problems as the modification of sterilisation conditions when coloured reaction products of sugars and proteins interfere with recovery, the adverse effect of medium modifications or the use of certain antifoam agents which may cause fouling when adsorption and membrane processes are used for the first recovery steps can often easily be solved by compromises convenient to both parties.

When dealing with processes involving recombinant microorganisms the whole procedure must be considered as a single entity. The future for chemical engineers associated with the scaling-up of recovery processes may be brighter if the proposal of Alan Michaels (1984) in *Chemical Engineering Progress* were to be readily accepted by life scientists. Under the heading 'Adapting modern biology to industrial practice' in the section 'The life science/engineering interface' he writes: 'Life scientists must become aware of and be more responsive to the problems of bioproduct isolation and recovery from cell cultures and fermentation broths; then they may be able to use their gene-cloning capabilities to assist the chemical engineer by generating cell lines which will facilitate rather than frustrate product recovery'.

ACKNOWLEDGEMENTS

The author thanks Mr J. A. L. Auden for his critical reading of the manuscript and for improving the English style.

REFERENCES

Bartels, Ch. R., Kleimann, G., Korzun, J. N. & Irish, D. B. (1958) A novel ion exchange method for the isolation of streptomycin. *Chem. Eng. Progr,* **54**, No. 8, 49–51.

Baum, R., (1984) Biotechnology scale-up: How big a problem? *Chem. Eng. News* **62**, No. 16, 13.

Burgoyne, R. F., Bowles, D. K. & Heckendorf, A., (1984). Rapid scale-up of peptide purification by preparative HPLC. *Am. Biotechnol. Lab.* March.

Chem. Eng. News (1968) **46**, No. 14, 62–63. Liquid chromatography looms as full scale separation method for production uses.

Chem. Proc. (1983) November. System tackles separations at 99% purity, 0.2–15 kg/h.

Ek, L. (1968) Gel filtration – a unit operation. *Proc. Biochem.* **3**, No. 9, 25–28.

German patent 2,306,210. Device and procedure for maintaining a low liquid level over a variable bed height ion exchanger (in German).

Gerstenberg, H., Sittig, W. & Zept, K. H. (1980) Processing of fermentation products. *Chem. Eng. Tech.* **52**, 19–31. *Ger. Chem. Eng.* **3**, 313–327.

Hoffman, J., Hermansky, M., Parizkova, H. & Doskocil, J. (1957) Isolation of streptomycin using a carboxylic ion exchanger (in German). *Chem Techn.*, **9**, No. 3, 151–155.

Janson, J.-C. & Dunnill, P. (1974) *Factors affecting scale-up of chromatography. Industrial Aspects of Biochemistry*. New York, American Elsevier, North Holland, pp. 81–105.

Martinola, F. & Siegers, G. (1979) Ion exchange – a large scale separation process (in German). *Verfahrenstechnik* **13**, No. 1, 32–35.

Michaels, A. (1984) Adapting modern biotechnology to industrial practice. *Chem. Eng. Progr.* **80**, No. 6, 19–25.

Pirotta, M. (1982) Amberlite ER 180 – a new styrene–divinylbenezene adsorbent specifically designed for industrial chromatography and particularly for the extraction of cephalosporin C. *Die Angewandte Makromolekulare Chemie* **109/110**, 197–214.

Shih, C. K., Snavely, C. M., Molnar, T. E., Meyer, J. L., Caldwell, W. B. & Paul, E. L. (1983) Large scale liquid chromatography system. *Chem. Eng. Progr.* **79**, No. 10, 53–57.

Strobel, G.-J. (1982) Large scale purification of biomolecules (in German). *Chemie Technik* **11**, 1354–1357.

US Patent 2,528,188. Two tower process for recovery of streptomycin employing cation exchange resins.

US Patent 3,250,058. Method of and apparatus for chromatographic separations.

US Patent 3,522,248. Process for the isolation of cephalosporin C.

US Patent 3,539,503. Large scale column chromatography and apparatus therefor.

US Patent 3,814,253. Liquid control apparatus for chromatography columns.

Voser, W. (1982) Isolation of hydrophilic fermentation products by adsorption chromatography. *J. Chem. Tech. Biotechnol.* **32**, 109–118.

9

Factors important in the design of fixed-bed adsorption processes for the purification of proteins

Howard Allaker Chase
Department of Chemical Engineering, University of Cambridge,
Pembroke Street, Cambridge CB2 3RA, UK

9.1 INTRODUCTION

Fixed-bed adsorption processes are important unit operations currently being used for the isolation and purification of proteins (Janson & Hedman 1982). A satisfactory procedure using fixed-bed adsorption can easily be developed on a laboratory scale and there is normally little economic incentive to optimise the process. However, when being considered for use on the industrial scale, economic considerations begin to dictate that such a process must be optimised. The aim of this chapter is to discuss factors that should be taken into consideration in the design of efficient, optimised procedures.

A number of biochemical separation principles can constitute fixed-bed adsorption processes. They include: ion exchange chromatography in which polyvalent, charged proteins bind to appropriately charged ion exchangers; affinity chromatography which exploits the specific recognition interactions that can occur between proteins and immobilised ligands; and a more general class normally referred to simply as adsorption chromatography where the mechanism of adsorption is often uncertain. Examples of the latter class include the use of hydroxyapatite where adsorption is due to hydrophilic interactions, and other non-charged adsorbents where hydrophobic interactions are implicated in the adsorption process. Gel permeation chromatography (gel filtration) is not an adsorption process and will not be considered further here.

The use of fixed-bed adsorption as a unit operation in the downstream processing of proteins has a number of advantages. Unlike many other protein separation methods, it is relatively easily scaled up from an optimised laboratory scale process, the separation protocol can be automated and high degrees of purification can often be achieved in a single-step process (Chase 1984a; Janson 1984). As is discussed later, configurations can be proposed that are capable of semi-continuous processing. On the other hand, fixed-bed adsorption processes do also have some disadvantages. The kinetics that describe the interaction between molecules immobilised to a solid phase and large molecules in solution may be slow and unfavourable (Lowe 1984). Indeed, there is considerable current interest in the use of aqueous two-phase extraction systems, with greatly improved kinetics, for the isolation of proteins (Mattiasson 1983). In addition, some adsorbents, and in particular those used in affinity chromatography, may be very expensive and the adsorbent in a large, industrial-scale bed can represent a considerable financial investment. Traditional fixed-bed methods are only suitable for solutions that have been rendered devoid of particulate matter and truely continuous separation processes may be difficult to achieve. However, fixed-bed adsorption remains a popular technique for large-scale processes and will now be considered in greater detail.

A fixed-bed adsorption process consists of a number of separate stages or phases of operation. These stages are illustrated in Fig. 9.1 for affinity chroma-

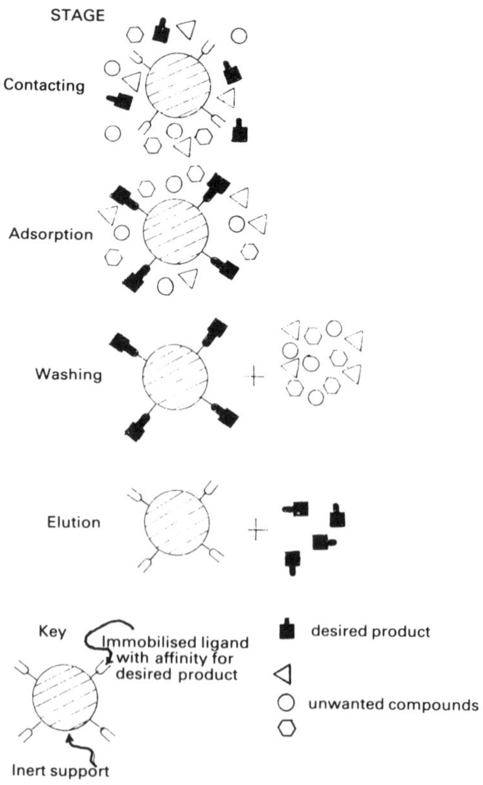

Fig. 9.1 – The various stages of an affinity chromatography separation.

tography. Initially the adsorbent is contacted with the crude mixture of proteins and during the adsorption stage only those proteins with an affinity for the adsorbent become bound to the solid phase. Proteins that are not adsorbed, and those that bind non-specifically are removed during the washing phase. The proteins adsorbed specifically are removed from the adsorbent during the elution phase after which the adsorbent can be re-equilibrated with an appropriate buffer solution and used in a further cycle of operation. The ability to be able to use the adsorbent for many cycles of operation is an important consideration in the financial feasibility of the process (Janson 1984). Similar stages of operation are also used in ion exchange and adsorption chromatography, but as the adsorbents are less selective, many different proteins may become bound during the adsorption stage and are not separated from each other until the elution stage of the process. The nature of the interactions that occur between proteins and the various adsorbents and the resultant selectivities shown by the adsorbents are considered further in Table 9.1.

Table 9.1 – The nature of the interactions between proteins and adsorbents.

The table gives details for a number of types of adsorption systems as to whether the adsorbent has unique binding sites for proteins of 'defined' structure or whether binding is to a random orientation of groups constituting an 'undefined' binding site. Indication is also given as to the specificity of the adsorbent, i.e. the number of proteins in a crude mixture that might be expected to bind to the adsorbent via the primary mechanism of adsorption.

Adsorbent	Nature of binding site	Number of specific interactions
Immobilised antibody	Defined	1 (antigen)
Immobilised substrate analogue	Defined	1 or more (enzymes that have a binding site for analogue)
Immobilised dye	Defined/ undefined	Several (specific interactions and secondary ion exchange/hydrophobic effects)
Ion exchanger	Undefined	Many (any proteins with an appropriate charge)
'Other' adsorbents	Undefined	Probably many (depending on the mechanism of adsorption)

9.2 EQUILIBRIUM INTERACTION MODEL

A number of authors have considered the interactions that occur between proteins and adsorbents in terms of a simple equilibrium of the following form (Graves & Wu 1974; Wankat 1974; Yang & Tsao 1982; Chase 1984b):

$$\underset{\text{protein}}{\mathrm{P}} \;+\; \underset{\text{adsorbent}}{\mathrm{A}} \;\underset{k_2}{\overset{k_1}{\rightleftharpoons}}\; \underset{\text{adsorbed complex}}{\mathrm{P.A}} \tag{1}$$

Protein (P) and adsorbent (A) react together reversibly to form a complex in which the protein is bound to the solid phase. The forward and reverse stages of the interaction are characterised by overall rate constants k_1 and k_2 respectively. These rate constants contain contributions from the various types of resistance to mass transfer that can occur, as well as contributions from the chemical events involved in the formation and breakdown of the adsorbed complex. Although the values of k_1 and k_2 may not actually be constant, it has been assumed for simplicity that an average overall value can, for most purposes, satisfactorily describe the situation (Chase 1984b). The adsorption interaction can be characterised by a dissociation constant $K = k_2/k_1$ with the units of concentration.

In order to achieve efficient adsorption of the desired protein, fast values of the adsorption rate constant k_1 and slow values of the desorption rate constant k_2 are required with a consequent small value of K. The equilibrium reaction shown in (1) will thus lie far over on the right-hand side and good adsorption will be achieved. Other proteins that are not required to bind to the adsorbent should have their interactions characterised by slow k_1 values and fast k_2 values, i.e. large values of K. Similar values of the rate constants describing the interactions during the adsorption stage are also desirable during the washing stage to ensure that the desired protein remains bound to the adsorbent whilst contaminants are removed. For elution, it is desirable that the adsorbed protein is removed from the adsorbent in the minimum amount of time and this requires a low value of k_1 and a high value of k_2. Hence during elution, the dissociation constant should have a large value resulting in a shift of the equilibrium position back to the left-hand side with release of the adsorbed protein.

One of the consequences of describing the adsorption process in terms of a simple equilibrium reaction is that the capacity of the adsorbent will depend on the concentration of protein in the soluble phase. The equilibrium capacity of the adsorbent (q) at equilibrium with a protein concentration (P) is given by (Chase 1984b):

$$q = \frac{q_{\mathrm{m}} \cdot P}{K + P} \tag{2}$$

where q_{m} is the maximum capacity of the adsorbent and K, the dissociation constant of the interaction, is that concentration of protein at equilibrium which results in the adsorbent showing half its maximum capacity. Measurement of adsorption isotherms is an important first step in the characterisation of the interaction between protein and adsorbent. The values of K and q_{m} pertaining to a number of adsorption systems have been measured in small-scale batch experiments and the values for some representative systems based on Sepharose as the matrix material are shown in Table 9.2. The values of q_m give an indication of the maximum possible capacity of the adsorbents although it must be remembered that such high values will not be achieved under most operating conditions. The values of K give some indication as to what concentrations of adsorbate are needed to achieve capacities approaching the maximum value, q_m. If the adsorption stage is carried out with protein concentrations of the same order as the value of K or smaller, only a fraction of the maximum capacity of the adsorbent will be utilised. Only if the concentration of protein is far greater than the value of K will the adsorbent show a capacity approaching q_m and even this capacity will occur only in circumstances where the protein and adsorbent have contacted for a sufficient period for equilibrium to be reached.

Table 9.2 – Equilibrium capacities and dissociation constants for various adsorption systems.

The table shows the values of the maximum capacities (q_m) and dissociation constants (K_d) for a number of adsorption systems using Sepharose (Pharmacia Fine Chemicals) as the adsorbent matrix. The values were measured from adsorption isotherms determined in small-scale batch systems where various concentrations of protein and adsorbent were equilibrated (Chase 1984b; Chase 1984c). All parameters were measured in 50 mM Tris/HCl buffer pH 7.2 at 20°C and the values for q_m are expressed per millilitre of settled adsorbent in order that they can be related to bed volume.

	q_m		K_d	
System	mg cm^{-3}	mol cm^{-3}	mg cm^{-3}	M
Sepharose 4B-anti-β-galactosidase β-galactosidase	2.9	6.2×10^{-9}	6.5×10^{-3}	1.4×10^{-8}
Sepharose CL6B-Cibacron Blue Bovine serum albumin	14	2.1×10^{-5}	0.56	8.5×10^{-6}
Sepharose CL6B-DEAE Bovine serum albumin	67	1.0×10^{-4}	0.091	1.4×10^{-6}

It must be stressed that the values of K and q_m only pertain to the circumstances under which they were measured and may be highly dependent on the particular conditions employed. As an illustration, the capacity of the ion exchanger DEAE–Sepharose CL-6B (Pharmacia Fine Chemicals) was measured at a number of pH values (Fig. 9.2). The values of K and q_m obtained from the

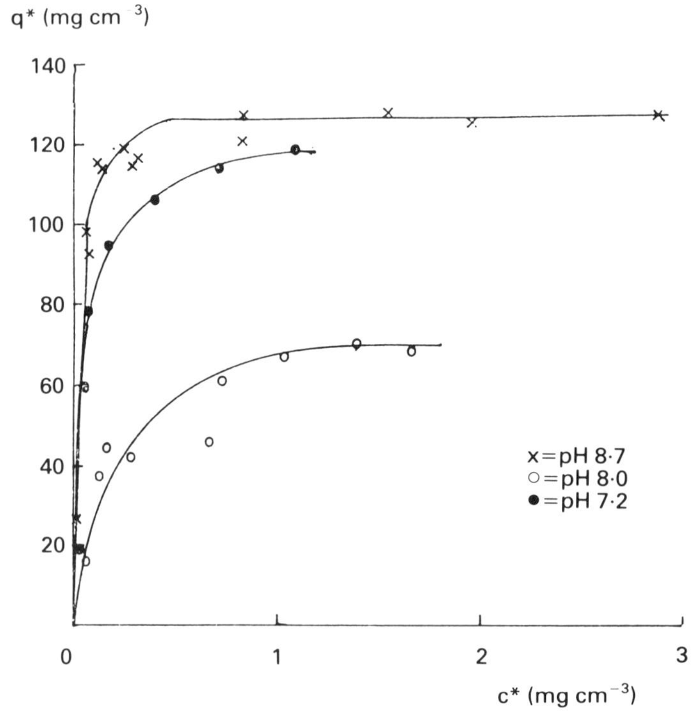

Fig. 9.2 – The effect of pH on the binding of bovine serum albumin to DEAE–Sepharose CL-6B. Solutions of bovine serum albumin in 50 mM Tris/HCl buffer at various pH values were equilibrated with various amounts of DEAE-Sepharose CL-6B. The concentration of protein in solution at equilibrium was determined by measuring the optical density of the liquid phase at 280 nm and the amount of protein bound per millilitre of settled adsorbent calculated (q^*).

adsorption isotherms are shown in Table 9.3, which also illustrates the large influence conferred by the nature of the protein on the values obtained (Chase 1984c). Bovine serum albumin, with a low iso-electric point, bound well to the positively charged ion exchanger whereas lysozyme, with a high iso-electric point was not bound at all. Myoglobin, with an iso-electric point very close to the pH of the buffer used for the experiments was found to bind weakly to the ion exchanger and accurate values for K and q_m could not be determined as the adsorbent was not saturated even at concentrations of myoglobin as high as 7 mg cm^{-3}.

The simple equilibrium model proposed for adsorption behaviour thus appears to agree well with the experimental data obtained. It must be remembered, however, that its application to adsorption systems where more than one protein can bind to the adsorbent will result in additional complexities. Under such circumstances, which occur for example in ion exchange chromatography, the various proteins present will compete for the available adsorption sites. During prolonged loading, the more tightly adsorbed proteins will displace loosely held

Table 9.3 – Parameters describing the interaction of proteins with the ion exchanger DEAE–Sepharose CL-6B.

Values of the maximum capacity (q_m) of DEAE–Sepharose CL-6B for various proteins at a range of pH values were determined from isotherms measured as described in Fig. 9.2. The table also shows the iso-electric point (pI) and relative molecular mass (M_r) of the proteins.

Protein	M_r	pI	q_m (mg cm^{-3})	K (mg cm^{-3})	pH
Albumin	69 000	5.1	134	0.017	8.7
			112	0.025	8.0
			67	0.091	7.2
Myoglobin	17 800	7.3	?	>7	8.7
			~0	–	7.2
Lysozyme	23 200	>11	0	–	8.7
			0	–	7.2

ones (Chase 1984c). We have had some success at applying a more complicated equilibrium model to account for the observed behaviour of mixtures of two or more proteins using knowledge of the kinetics of the interactions measured with the individual proteins. The equilibrium model provides a useful basis for predicting the performance of the various stages of a fixed-bed separation procedure.

9.3 DESIGN OF ADSORPTION STAGE

Although the measurement of equilibrium maximum adsorption capacities enables some prediction of the capacity of adsorbents during fixed-bed operation, the actual capacities observed under these circumstances are likely to be considerably less than expected (Chase 1984b). This is because it is very unlikely that the duration of the adsorption stage will be continued until equilibrium between the adsorbent and the adsorbate has been reached. This point is explained in greater detail by reference to the background curves (frontal analysis) shown in Fig. 9.3. Figure 9.3(a) shows that during fixed-bed operation, the equilibrium bed capacity is reached only after a considerable amount of the adsorbate applied to the bed has flowed straight through without being retained. Such a loss of adsorbate would be unacceptable in a separation process and loading of the bed would be terminated either before the level of adsorbate in the outlet stream rose above a predetermined acceptable value as illustrated in Fig. 9.3(b), or when the proportion of the total adsorbate applied that had not been retained

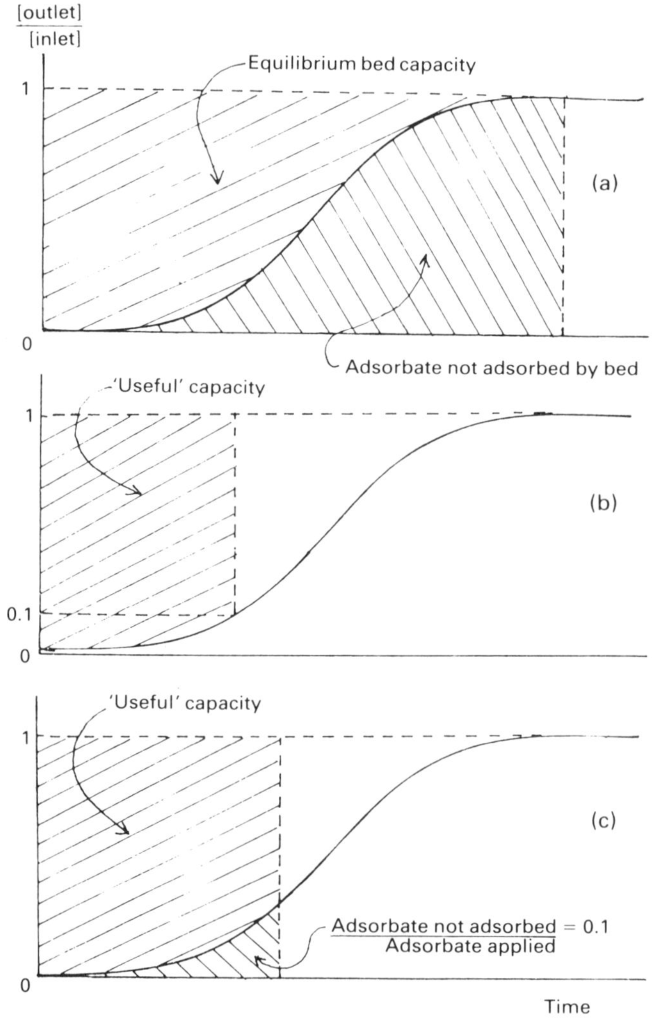

Fig. 9.3 – Adsorption breakthrough curves for fixed-bed operation and their effect on bed capacity. The solid lines show the ratio of the concentration of protein in the outlet of the bed to the concentration in the inlet (breakthrough curves). When this ratio has risen to 1, the bed is at equilibrium with the inlet stream and no further adsorption occurs. (a) Shows that the equilibrium bed capacity is only reached after a considerable amount of adsorbate has passed through the bed without being adsorbed. (b) Shows the 'useful' capacity of the bed if loading is terminated when the concentration ratio reaches an arbitrary value of 0.1. (c) Shows the useful capacity if loading is terminated when the ratio of the amount of adsorbate not retained by the bed to the total amount of adsorbate applied by that time is arbitrarily 0.1.

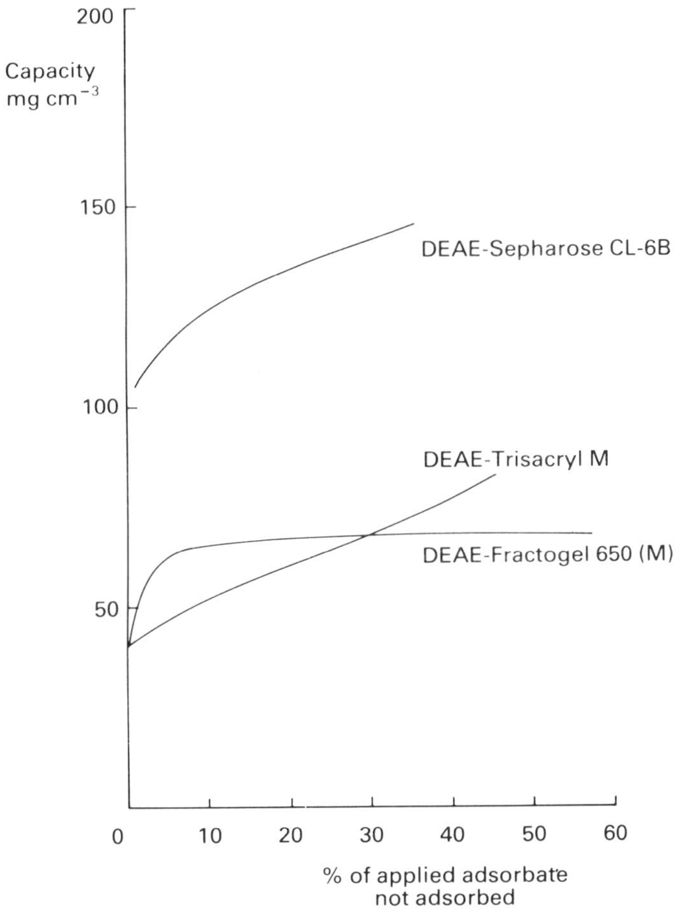

Fig. 9.4 – Variation of the capacity of ion exchangers with the extent of adsorbate breakthrough. A solution of bovine serum albumin (5 mg cm^{-3} in 50 mM Tris/HCl buffer pH 8.7) was applied to beds (26 mm length × 16 mm diameter) of various ion exchangers at 20°C at a linear flow rate of 10 mm min^{-1}. Breakthrough curves were measured by monitoring the optical density of the fluid in the bed outlet at 280 nm. At a number of time points, the amount of bovine serum albumin retained by the bed per millilitre of adsorbent was calculated together with the amount of applied adsorbate that had broken through the bed without being retained (see Fig. 9.3 (c) for details).

by the bed rose to an unacceptable value (Fig. 9.3(c)). The latter is a more efficient criterion as to when the working capacity of the bed has been utilised. Whichever of the two criteria for termination are adopted, it is clear that the 'useful' capacity of the bed is less than the equilibrium capacity. The useful capacity of the bed will depend on the degree of flow-through of adsorbate that is considered acceptable in the separation. Figure 9.4 shows the variation of useful capacity of an adsorbent with the degree of breakthrough of adsorbate for three ion exchangers in which the DEAE function is attached to different matrix materials. The characteristics of the adsorbents are very different, with the ion exchanger based on Sepharose CL-6B showing the highest capacity per unit bed volume at all degrees of breakthrough. The exchanger based on the Fractogel 650 (M) matrix (E. Merck) showed a flat characteristic indicating that the equilibrium capacity was reached quickly (i.e. the breakthrough curve was sharp). The ion exchanger based on Trisacryl M (L.K.B. Instruments Ltd) had a capacity that almost doubled as the degree of breakthrough increased (i.e. had a shallow breakthrough curve). These results show that although various matrix materials have been developed recently to withstand the mechanical stresses associated with operation at high flow rates, in practice the kinetics of the adsorption processes may require the use of substantially lower flow rates in order to be able to achieve high useful bed capacities.

It is obvious therefore that the sharpness of the breakthrough curve plays a critical role in the determination of the useful capacity of an adsorbent under a particular set of operating conditions. The sharper the shape of the breakthrough curve, the closer the useful capacity will approach the equilibrium capacity at low degrees of breakthrough. Factors that influence the shape and position of the breakthrough curve have been considered in detail (Chase 1984b) and include the flow rate through the bed, the bed size, the concentration of adsorbate in the inlet stream, the value of the dissociation constant K describing the adsorption reaction, the rate constants of the adsorption/desorption reactions and the existence of an uneven distribution of flow across the bed. The importance of obtaining an even flow distribution in large-scale beds has been emphasised in detail in Chapter 8 by Voser and Walliser. In general, shallow breakthrough curves are likely to be inefficient but can normally be made sharper by a combination of lowering the flow rate and/or increasing the size of the bed. Neither of these actions is particularly desirable in large-scale applications where processing in the minimum time with the minimum size of bed is actively sought. In some circumstances when the concentration of adsorbate is not large compared to the value of the dissociation constant of the adsorption reaction, adsorption performance can be improved by prior concentration of the adsorbate solution.

Throughout this discussion of adsorption performance, it has been assumed that it is desirable to load the bed to near its maximum capacity. There may, however, be a number of applications where such a high degree of loading is not desirable – in particular during ion exchange processes where it is only possible to separate certain proteins from one another during the elution stage. In order to obtain good resolution of the eluted proteins under such circumstances, it would be necessary to load the bed only to a limited degree so that the remaining unutilised capacity of the bed can participate in resolving the eluted proteins. In such cases, the adsorption stage of the process has to be optimised in conjunction with the performance of the elution stage. However, conditions that favour sharp breakthrough curves will also result in better sub-saturating loading protocols leading to sharper resolution upon elution.

9.4 DESIGN OF WASHING AND ELUTION STAGES

The aims for a successful washing stage in a fixed-bed adsorption procedure are to obtain maximum removal of contaminants bound non-specifically to the bed consistent with minimum removal of the specifically bound product. A washing stage is of particular importance in affinity chromatography as proteins often adsorb to the matrix material or other parts of the attached ligands. These contaminants have to be removed before elution otherwise their presence in the eluate will lower the degree of purification achieved. For less specific adsorption systems, the inclusion of a washing stage is less important as the adsorbed proteins are separated from each other on elution. Hence the following discussion of the washing stage is mostly appropriate only to affinity chromatographic separations.

At the end of the adsorption phase, there will be crude starting material in the voids of the bed. One of the functions of the washing stage is to remove material in the voids of the bed and one bed volume of wash buffer should be sufficient. The effectiveness of the wash buffer at removing contaminants that are bound non-specifically to the adsorbent will depend on the strength by which they are bound, with the more strongly adsorbed contaminants being removed more slowly. Hence, it is usual to use a wash buffer with different physical properties (e.g. ionic strength, dielectric constant, pH) to the starting material so that the affinity of the contaminants for the adsorbent is decreased and thus the contaminants are, in effect, eluted. As an example, when using immobilised antibodies as affinity adsorbents, we routinely carry out a washing stage with 1M NaCl in order to remove any proteins bound to the adsorbent by ion exchange effects. Solutions of detergents may help to remove contaminants that are bound by hydrophobic interactions (Eveleigh & Levy, 1977). Assuming that the cost of washing buffer is not great, it is better to wash the bed at a high flow rate to ensure greatest removal of contaminants in a given period of time. The use of low flow rates will ensure better removal with respect to the number of bed volumes of washing buffer used, but will lengthen the period of the washing stage.

The extent of loss of the specifically adsorbed proteins during the washing stage will depend on a number of factors and in particular to what degree the bed has been loaded to in the adsorption stage. If the bed has been loaded to its capacity, there is the danger that when specifically adsorbed material desorbs from the adsorbent during the wash (because of the reversible nature of the adsorption reaction), it may not become bound again before emerging from the bed exit. If the bed is not highly loaded, there is a much greater chance that the protein will re-adsorb to vacant sites on the bed. The extent of loss will also depend on how strongly the protein is adsorbed to the bed and to whether the wash buffer lowers the affinity of the protein for the specific sites on the bed. Unfortunately, most washing protocols that are effective at removing contaminants will also promote the loss of the desired product. Hence the design of the washing stage has to be a compromise between the desired purity and yield of the product.

The aims of the elution stage are to achieve efficient removal of the adsorbed proteins and, where the less specific adsorbents are used, to separate the adsorbed proteins during elution. If only a single protein is bound to the adsorbent or if it is not desired to separate the adsorbed species, elution with a single-step change of eluent is normally sufficient. The role of the eluent is to lower the affinity of the adsorbed protein for the adsorbent. This may be achieved in a specific manner in affinity chromatography by including the soluble form of the immobilised ligand in the irrigating buffer to achieve competition with the immobilised ligands for the binding site on the protein. Such an approach may be impracticable in some applications because of the difficulties involved in the subsequent separation of the eluted protein from the soluble ligand. In addition, the use of soluble ligand may be too expensive in large-scale applications. Alternatively, non-specific elution requires the use of an eluent that causes the strength of the binding between the protein and the adsorbent to be reduced. As examples, the use of high ionic strength or an appropriate change in pH results in the elution of proteins from ion exchangers (Chase 1984c). Similarly, the use of mild denaturants will enable the elution of proteins from immunoadsorbents by altering the three-dimensional structure of the adsorbed protein or the immobilised antibody such that the affinity between them is reduced (Kristiansen 1978). Ideally, eluents should result in high values of the rate constant k_2 describing the breakdown of the adsorbed complex in order to get rapid release of the protein and low values of k_1 to discourage the reformation of a complex (Lowe & Dean 1974). If it is desired to separate proteins during elution, it is necessary to alter the strengths by which they are bound to the adsorbent in a selective manner. This can be achieved by the use of multiple step changes or continuous gradient changes in the concentrations of a non-specific eluent. Alternatively, separation can be achieved by using a sequence of specific eluents as has been used to separate a number of enzymes adsorbed to an immobilised dye as described by Lowe and Stead in Chapter 10.

The use of low flow rates during elution will result in the product being eluted in the minimum volume and hence at the highest concentration. This is because the protein will have been given a long time to desorb from the bed. Indeed, better elution may be achieved by halting the flow through the bed from time to time to promote maximum desorption in the minimum volume. If it is desirable that the product is eluted in the minimum time regardless of volume because the protein denatures in the presence of eluent, high flow rates will result in the bed being contacted continuously with eluent devoid of desorbed protein and will favour rapid desorption. Gradient elutions are usually carried out using a constant, slow flow rate to achieve maximum resolution of the desorbed proteins. Gradient elutions are performed using a direction of flow the same as that used for loading the bed so that the full length of the bed can be used to resolve the eluting proteins. When step changes are used in elution, it is generally more efficient to reverse the direction of flow through the bed (Kristiansen 1978). This is to avoid having to flow the eluted protein through the remainder of the bed which would increase the chance that the protein readsorb to the bed and ultimately would result in an increased elution volume. Figure 9.5 shows the results when an affinity adsorbent was eluted in both directions with a weak eluent and it is clear that when reverse flow was used for elution, the product was eluted in a smaller volume. If the bed had been loaded to near its capacity, the direction of elution would make very little difference as there would be an even distribution of adsorbed protein throughout the bed.

The expected shape of the eluted peak is asymmetric with a sharp front and a trailing tail (Graves & Wu 1974). It is important to ensure that all the adsorbed protein is removed from the bed during elution in order that the capacity of the adsorbent is not reduced during the subsequent cycle of operation. Removal of last traces of adsorbed protein is made difficult by the trailing tail of the eluted peak and may result in substantial dilution of the product unless separate fractions are collected.

9.5 MONITORING AND AUTOMATIC CONTROL

The industrial use of fixed-bed adsorption separations requires the development of automatic control systems and monitoring in order to achieve reliable unattended operation. A suitable control system basically consists of a set of valves to regulate the application of fluid streams to the bed (starting material, washes, eluents, etc.) and the collection of fluid from the bed (waste streams, product, non-adsorbed material) together with a set of instruments to provide a record of the separation. One simple method of control is to change the positions of the valves on the basis of a predetermined time sequence. Although such a system is simple to implement (Eveleigh 1982), for optimum efficiency to be achieved, the separation must be reproducible from cycle to cycle and conditions such as the composition of the starting material must be constant. Otherwise, abnormally low concentrations of adsorbate in the starting material will result in under-utilisation of bed capacity whereas abnormally high adsorbate concentrations may exceed the capacity of the bed if the adsorption stage is carried out for a constant period. A more efficient mode of operation can be achieved by using interactive control where the performance of the bed is monitored on-line (by analysing the inlet and outlet streams of the bed) and the results used for the control of fluids to and from the bed (Chase 1984d). Such a system would potentially be able to cope with varying inlet concentrations of adsorbate and any reduction in the capacity of the adsorbent that may occur during repetitive operation.

However, interactive systems need suitable monitoring equipment to follow the progress of the process and this may not always be available. Continuous flow spectrophotometers are commonly used to measure the overall concentration of protein. However, they cannot detect the level of a particular protein in the presence of other proteins and although such instruments are useful for following the washing and elution stages of the process, they have limited use

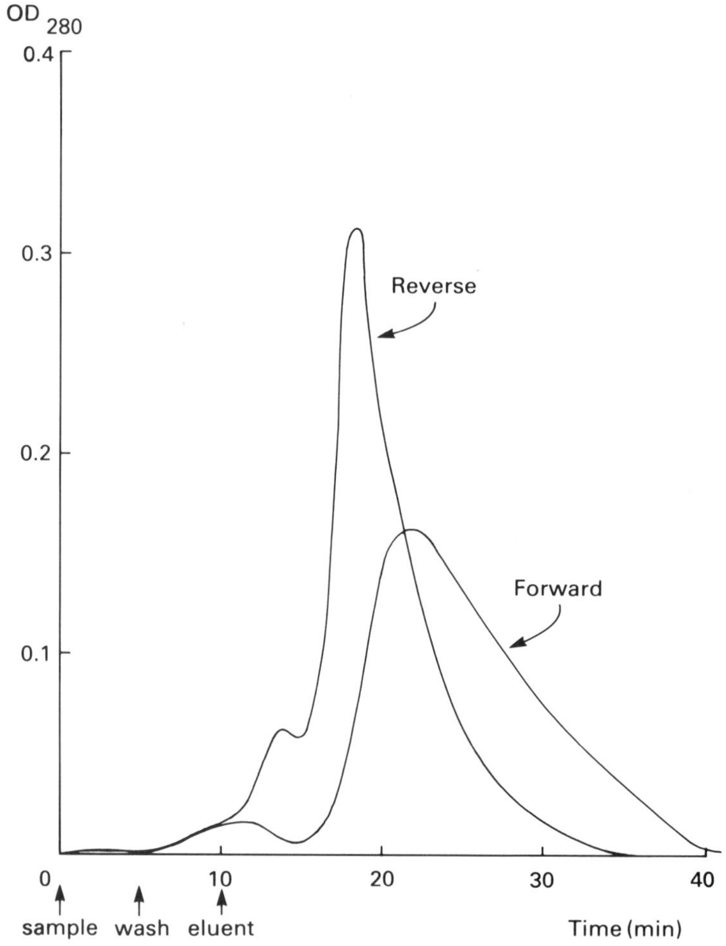

Fig. 9.5 – Effect of the direction of fluid flow during elution. Lysozyme (1 mg cm^{-3} in 50 mM Tris/HCl, pH 7.2) was applied to a 50 mm length × 16 mm diameter bed of Cibacron Blue–Sepharose CL-6B (Pharmacia Fine Chemicals) for 5 min at a volumetric flow rate of 1 cm^3 min^{-1}. The lysozyme was washed into the bed with a further 5 cm^3 of the Tris buffer. Elution with 0.2 M KCl in Tris buffer was carried out in one experiment in the same (i.e. forward) direction of flow and in a subsequent run after reversing the direction of flow through the bed. The optical density at 280 nm was measured with a continuous flow ultraviolet spectrophotometer.

in following the progress of the adsorption stage. Continuous flow pH or conductivity meters are useful for following the changes of pH or ionic strength during washing, elution and re-equilibration. For interactive systems to be effective, special methods are needed to measure the levels of specific proteins in protein mixtures. Unless the protein of interest has some unique property (e.g. it has enzymic activity, is radioactively labelled, or has an unusual spectroscopic adsorption band), breakthrough of the protein being adsorbed by the bed cannot be followed on-line. The widespread use of interactive control will have to await the development of biosensors capable of detecting the levels of particular proteins; such biosensors should become generally available within the next few years (Gronow 1984). Figure 9.6 shows the outline of a microcomputer-controlled automatic chromatography control system that we have developed. The configuration shown includes an on-line gel filtration system for the rapid removal of eluent from the eluted protein. We have found this facility particularly useful in our studies of elution from immunoadsorbents where the eluents are often denaturants and their rapid removal is desirable (Chase 1984d).

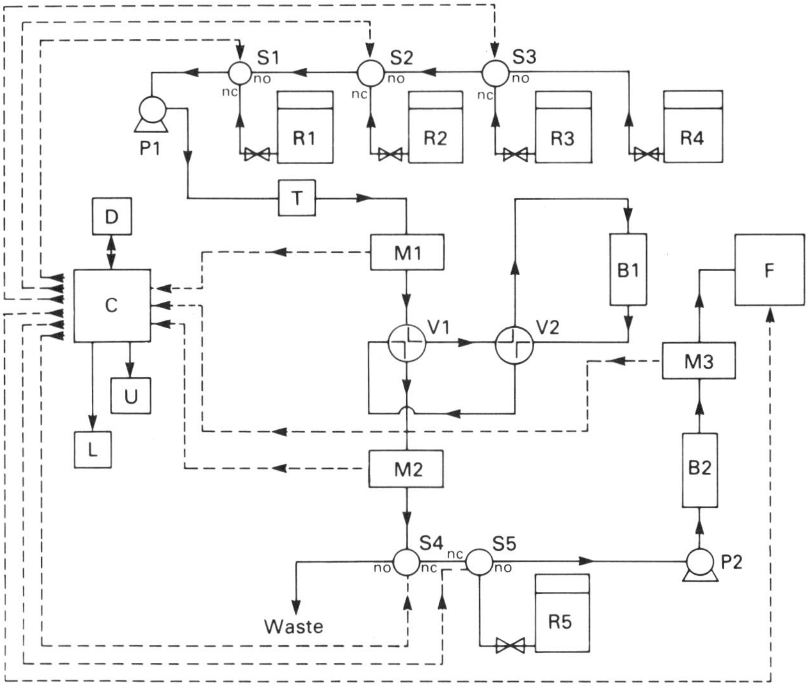

Fig. 9.6 – Microcomputer-based system for the automatic control of fixed bed separations. B fixed bed, B1 = immunoadsorbent, B2 = gel filtration; C computer; D data storage unit; F fraction collector; L line printer; M flow spectrophotometer (280 nm); P peristaltic pump, P1 and P2 are two identical heads on the same pump; R reservoir, R1 = crude starting material, R2 = wash buffer, R3 = eluent solution, R4 = bed re-equilibribration buffer, R5 = gel filtration buffer; T bubble trap; S solenoid valve (no = port normally open, nc = port normally closed); U visual display unit; V manually operated valve.

There is often a desire, in the development of industrial processes for the downstream processing of proteins, to construct systems capable of continuous rather than batch processing (Lowe 1984). Two types of continuous processing can be identified: truly continuous systems which continuously process the input stream and produce a continuous level of the eluted product, and semi-continuous systems in which the input stream is processed continuously but the product is eluted in periodic bursts.

One effective way of achieving truly continuous processing in adsorption systems is to have countercurrent movement of the adsorbent. Such systems can be achieved by having physical movement of the absorbent and moving-bed systems are already in use for certain ion exchange processes other than protein separation (Van der Meer *et al.* 1984). The physical nature of the adsorbents currently in use for protein isolation are unlikely to have suitable mechanical properties to withstand the considerable matrix attrition that can occur in such systems. Alternatively, countercurrent operation can be achieved in stationary-bed systems by continuously altering the position of the feed point in the bed (Barker *et al.* 1978). Such systems involve complex valving arrangements, particularly in situations where true chromatography does not occur during the separation and proteins have to be removed from the bed with eluents.

We have developed a semi-continuous protein processing system based on the use of two stationary beds with an appropriate fluid switching system. When one bed has been loaded with protein, the inlet stream is diverted to the second bed whilst the first bed is washed and eluted. Provided that the time period required for washing and elution is less than that for loading, the first bed will again be ready for the adsorption stage by the time the second bed is loaded. The second bed is then washed and eluted whilst the first bed is again loading. The cyclic operation can then be repeated continuously. The periodic bursts of elution that occur in such a system are not in practice a problem as the concentration of eluted protein can be held reasonably constant by the introduction of a small, stirred holding tank before the next process in the overall recovery sequence. The outline of a simple system based on such principles is shown in Fig. 9.7 and its detailed operation will be described more fully elsewhere. The system can be automatically controlled by our microcomputer-based chromatography controller and promises to be a comparatively simple solution to the problem of achieving a continuous processing system.

9.6 OTHER FACTORS OF IMPORTANCE

The choice of the position of an adsorption process in a protein recovery flow-sheet is an important decision. If the process is inserted at an early stage of the recovery, there are likely to be large volumes of fluid to pass through the bed with the adsorbate present in the fluid at low concentration. In such situations, large beds will be needed so that the fluid can be processed in a reasonable time for a given bed residence time. The crude material at an early stage is likely to contain a large number of compounds some of which may result in bed fouling or

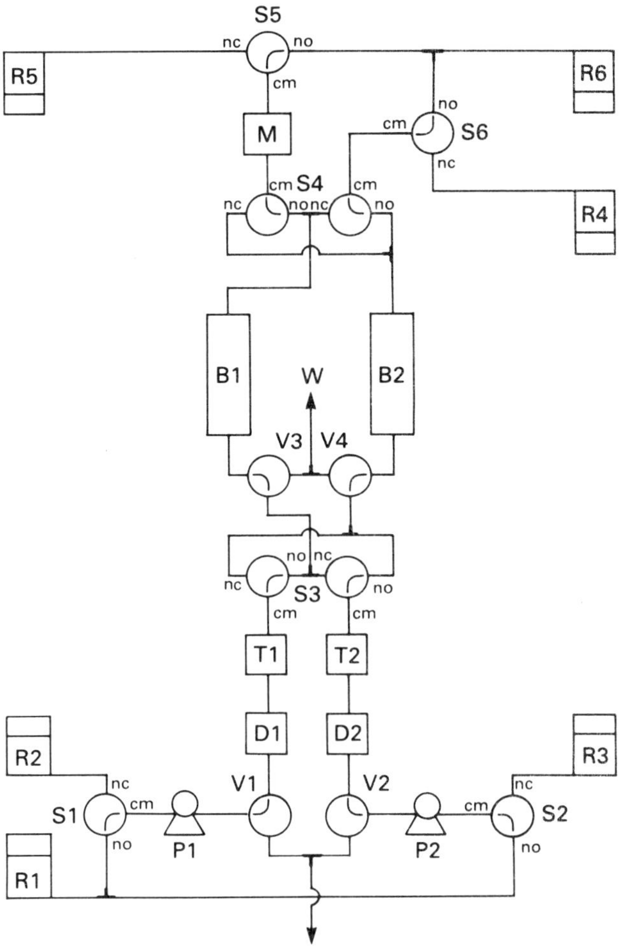

Fig. 9.7 – Outline of a two-bed system for semi-continuous processing. B1, B2 identical beds of adsorbent; D1, D2 error detection systems (detect over-pressure and fluid loss); M continuous flow ultraviolet spectrophotometer (280 nm); P1, P2 peristaltic pumps; R fluid reservoir, R1 = re-equilibration/wash buffer, R2 = starting material (from previous process), R3 = elution buffer; R4 = waste; R5 = product (to next stage); R6 = non-adsorbed material; S1–S6 electrically operated three-way solenoid valves; T1, T2 bubble traps; V1–V4 manually operated valves (used for system set-up); W waste stream. The valves S1–S6 are operated in a predetermined sequence by a minicomputer (as in Fig. 9.6) on the basis of the signals sent from the spectrophotometer M. The principle of the semi-continuous operation is explained in the text.

the inactivation of fragile ligands on the adosrbent. Such fouling and inactivation are not desirable particularly when expensive, highly specific affinity adsorbents are being used. With the less specific adsorbents, the range of proteins adsorbed will be greater and it will be more difficult to achieve good resolution and hence high degrees of purification. On the other hand, the inclusion of a specific adsorption separation process such as affinity chromatography at an early stage does allow the possibility of achieving a high degree of purification at a single step, thus reducing the complexity and expense involved in constructing a multiple-step purification procedure.

Reservation of the use of an adsorption separation process to a late stage in a purification procedure means that smaller, cheaper beds can be used, as the volumes of fluid will generally be less and the concentration of adsorbate higher. As there will be fewer components present at a late stage, greater purification can be achieved with the less specific adsorbents and there is consequently less risk of bed fouling and inactivation. However, such conservatism in the use of the process may result in considerable under-utilisation of the purifying power of a specific adsorption separation process. No general rules can be given for determining the best stage to include such a process and the decision can only be made by investigating the relative importance of the factors described above to the particular recovery process under investigation.

There are a number of other factors of importance in the design of large-scale fixed-bed separations that will now be mentioned briefly but not dwelt on. The establishment and maintenance of sterility during operation is important in order to preclude the presence of microorganisms or their products in the final preparation and to maximise the longevity of the adsorbent (Van Wezel & Van der Marel 1982). Similarly, the leakage of parts of the matrix or the attached ligands is undesirable, not only because of the resultant contamination of the product but also because such breakdown will give rise to loss of adsorbent performance (Chase 1984d). The choice of a suitable matrix material is important in order to achieve fast kinetics for the adsorption/desorption reactions, high adsorbent capacities (to minimise bed sizes) and good mechanical properties to resist compression and compaction in large beds (Cooney 1984). A number of factors are important in the choice of ligand used in forming the adsorbent. The various factors that should be considered include the degree of purification that can be achieved by the ligand, its longevity (i.e. the number of cycles of operation that it can be used for) and, last but not least, its overall expense.

REFERENCES

Barker, P. E., Ellison, F. J. & Hatt, B. W. (1978) In: Epton, R. (ed.), *Chromatography of Synthetic and Biological Polymers*, vol. 1. Chichester, Ellis Horwood, pp. 218–239.

Chase, H. A. (1984a) *J. Biotechnol.* **1**, 68–80.

Chase, H. A. (1984b) *J. Chromatogr.* **297**, 179–202.

Chase, H. A. (1984c) In: Naden, D. and Streat, M. (eds.), *Ion Exchange Tecnology.* Chichester, Ellis Horwood, pp. 400–406.
Chase, H. A. (1984d) *Chem. Eng. Sci.* **39**, 1099–1125.
Cooney, J. M. (1984) *Biotechnology* **2**, 41–55.
Eveleigh, J. W. (1982) *Anal. Chem. Symp. Ser.* **9**, 293–303.
Eveleigh, J. W. & Levy, D. E. (1977) *J. Solid-phase Biochem.* **2**, 45–78.
Graves, D. J. & Wu, Y. T. (1974) *Method. Enzymol.* **34**, 140–161.
Gronow, M. (1984) *Trend. Biochem. Sci.* **9**, 336–340.
Janson, J.-C. (1984) *Trend. Biotechnol.* **2**, 31–38.
Janson, J.-C. & Hedman, P. (1982) *Adv. Biochem. Engg.* **25**, 43–99.
Kristiansen, T. (1978) In: Hoffman-Ostenhof, O. (ed.), *Affinity Chromatography.* Oxford, Pergamon Press, pp. 191–206.
Lowe, C. R. (1984) *J. Biotechnol.* **1**, 3–12.
Lowe, C. R. & Dean, P. D. G. (1974) *Affinity Chromatography.* London, John Wiley.
Mattiasson, B. (1983) *Trend Biotechnol.* **1**, 16–20.
Van der Meer, A. P., Weve, D. N. M. M. & Wesselingh, J. A. (1984) In: Naden, D. and Streat, M. (eds.), *Ion Exchange Technology.* Chichester, Ellis Horwood, pp. 284–294.
Van Wezel, A. L. & Van der Marel, P. (1982) *Anal. Chem. Symp. Ser.* **9**, 283–292.
Wankat, P. C. (1974) *Anal. Chem.* **46**, 1400–1408.
Yang, C. & Tsao, G. T. (1982) *Adv. Biochem. Engg.* **25**, 1–18.

10

The use of reactive dyestuffs in the isolation of proteins

C. R. Lowe
The Biotechnology Centre, University of Cambridge, Downing Street, Cambridge CB2 3EF, UK, and

C. V. Stead
Imperial Chemical Industries (Organics Division) PLC, Hexagon House, Blackley, Manchester M9 3DA, UK

10.1 CHEMISTRY OF CHLOROTRIAZINE DYES (C. V. Stead)

There has always been a strong affinity of interests between biochemistry and dyestuffs chemistry and, indeed, the earliest biochemical research sprang from dyestuffs chemistry. This is not surprising since the coloration of a fabric relies on the exploitation of an interaction between a macromolecule, which in the case of cellulosic and animal fibres is a natural macromolecule, and a synthetic organic compound. Over the years the products of dyestuffs research have often been of value in biochemical applications and it is, therefore, entirely appropriate that the modern fibre-reactive dyes (Beech 1970; Venkataraman, 1972) should have a role to play within the biochemical domain.

Fibre-reactive dyes were introduced to overcome certain problems which had beset the dyeing of the cellulosic fibres, notably cotton. When cotton is dyed it is necessary to exploit some intereaction between the fibre and the dyestuff in order to ensure satisfactory resistance to desorption during subsequent washing of the fabric. This is a particularly stringent requirement in the case of cotton since the fibre is very hydrophilic, thus offering no automatic protection from the effects of water. Further, the dyeing process must use water as the medium of application of the dye since only water is economically acceptable.

For many years physical effects were relied on to achieve satisfactory anchoring of the dye to the fibre in order to ensure wash fastness. Two such effects were exploited. The first was to prepare an insoluble dye from water soluble precursors on the fibre and rely on trapping of insoluble dye aggregates within the fibre pores. This was the mode of retention of vat and azoic dyes. The second mechanism relied upon building a water soluble dye which contained a multiplicity of sites capable of hydrogen bonding with the fibre and the cumulative effect of many such bonds to retain the dye on the fibre. Whilst both mechanisms had and retain their place in the dyeing of cotton, neither was entirely satisfactory for all outlets and it had long been perceived that a method of attaching a dye to the fibre through a chemical bond would offer a desirable advance.

The problem posed was how to devise a dye which was sufficiently reactive to react with the hydroxyl groups of the cellulose molecule yet which could be applied from aqueous medium. Attempts started as early as 1894 to achieve such a mode of attachment but they were doomed to remain mere chemical exercises since they invariably ducked the problem and resorted to application from a non-aqueous medium. These could never succeed commercially and they remain in the literature as methods for the production of coloured cellulose derivatives having made no impact as technically acceptable dyeing methods. It was not until 1956 that the problem was satisfactorily solved, the first solution finding its inspiration in the chemistry of cyanuric chloride.

Cyanuric chloride (I) contains three chlorine atoms activated by their attachment to the triazine nucleus. Activation results from the participation of canonical forms such as (II) allied to polarisation of the carbon–chlorine bond (III). Although reasonably stable in cold aqueous suspension at neutral

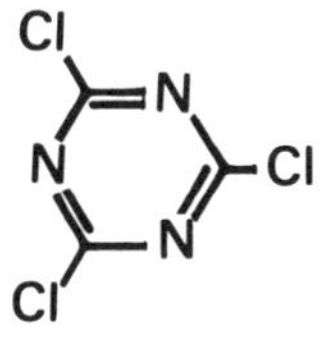

(I)

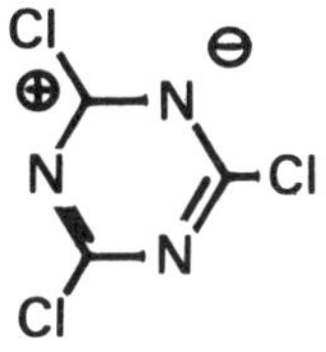

(II)

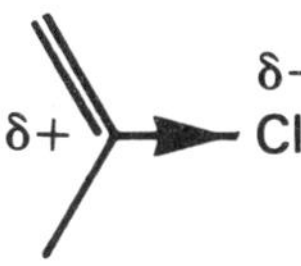

(III)

pH values ready reaction occurs with a suitable nucleophile. Such a nucleophile can be either a water soluble dye containing, for example, an amino group or a similarly substituted dyestuff intermediate, ready reaction occurring to produce a dichloro-*s*-triazinyl dyestuff (IV). Where an intermediate compound is used subsequent processing stages lead to conversion to the desired dye.

In the dichloro-*s*-triazinyl dye the electronic influences responsible for the reactivity of the chlorine atoms are muted by feedback of charge from the linking amino group (V), making the species quite stable in aqueous solutions at near-neutral pH values. This allows the dye to be isolated by precipitation with salt, filtration and drying. Buffer, usually a mixture of sodium dihydrogen and disodium monohydrogen phosphates, is added both to the solution and to the isolated dyestuff paste prior to drying in order to maintain the pH at about 7 (Horrobin 1963). The salt content is somewhat variable from batch to batch

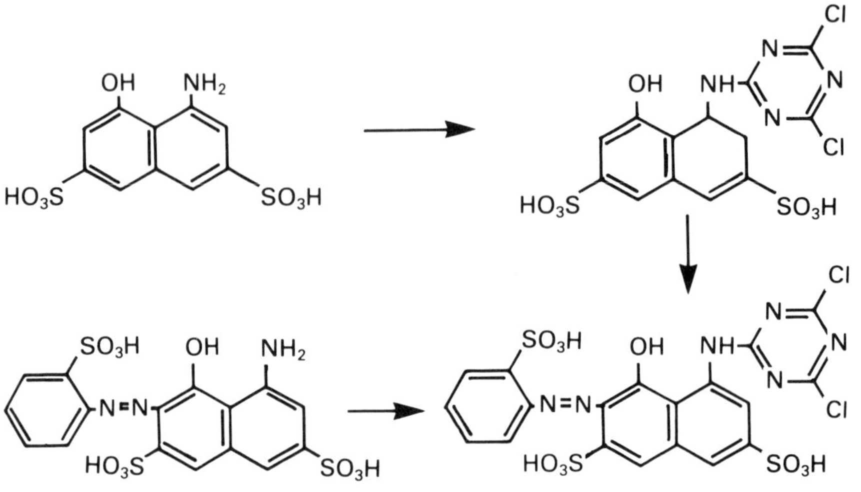

(IV)

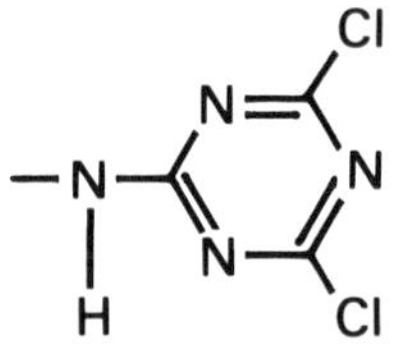

(V)

and to ensure reproducibility of effect additional diluent, usually salt, is added as required to standardise the effective agent content of the powder. Finally, a de-dusting agent, e.g. dodecylbenzene, is added to make the powder more clean to handle. Thus, the commercial dye powder contains, in addition to the dyestuffs, buffer, diluent and dedusting agent. Requests for dye samples from a manufacturer will normally be met by material of this quality. None of these adulterants appear to cause a problem in affinity chromatography applications.

In use the dye powder is dissolved in water and the cotton is immersed in the solution. Salt is added, causing the dye to be exhausted from the aqueous phase onto the fibre phase. About 30–45 minutes are required for this to be satisfactorily accomplished. Since all these physical changes take place at neutrality, no sensible amounts of nucleophiles being present, chemical reaction does not occur. When the dye has been satisfactorily adsorbed onto the fabric, the pH is raised by adding an alkali, often sodium carbonate. This generates the nucleophilic species required for reaction of the dye, which occurs smoothly at a temperature of about 40°C and over a period of about 45 minutes in the case of dichloro-*s*-triazine. Such reaction can proceed in one of two ways.

Reaction can occur with either ionised cellulose, fixing the dye to the fibre, or with a hydroxyl ion, in which case the dye is simply hydrolysed and lost to the dyeing process.

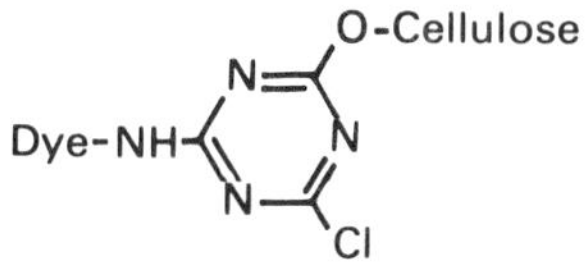

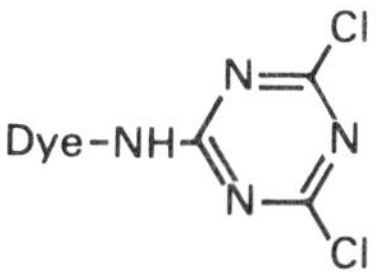

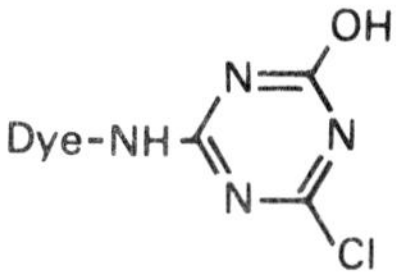

All other things being equal, the vast excess of water in the system would ensure that hydrolysis was the favoured reaction, causing the process to result in abject failure. Two factors intervene to alter the course of events. First, ionisation of the hydroxyl groups on the cellulose chain sets up an internal pH within the fibre which although lower than the dye bath pH is maintained primarily by the Cellulose—$O^{\ominus}$ ions, hydroxyl ions being repelled from the fibre interior. Second, the dye has been previously exhausted onto the fibre leading to the reactive dye finding itself in an environment where the ionised hydroxyl groups on the cellulose have preferential access to the dye. These effects combine to allow fixation to predominate over hydrolysis and enable the process to succeed (Ingamells *et al.* 1960). Of course, a certain amount of the dye does become hydrolysed and this is removed at the end of the dyeing process by thorough washing. The dyeing process thus described is known as exhaust dyeing. Many other methods of application have been designed for the application of reactive dyes, but all utilise the basic principles outlined above for their efficiency.

The parallels between this dyeing process and attachment of the reactive dye to any desired substrate (e.g. agarose) are clear. Again a preliminary period wherein the dye is physically adsorbed onto the substrate will be essential in achieving maximum fixation although, of course, in a biochemical application there will not be the same economic necessity to maximise dye usage. Nevertheless this adsorption period would be helpful in getting increased fixation of dye to the substrate and additionally might be helpful in getting an even reaction of dye throughout the substrate.

In passing it may be noted that should hydrolysis rather than fixation occur, the dye does not have a further chance to achieve attachment to the fibre by virtue of its remaining chlorine atom. This is because ionization of the acidic hydroxyl group (VI) now located on the triazine ring results in feedback of negative charge into the heterocyclic ring causing total deactivation of the remaining chlorine atom.

So far, we have concentrated on the dichloro-*s*-triazinyl type of dye which are found in the PROCION MX range of reactive dyes (ICI patents). A typical dye name would be PROCION Red MX-5B, the terminal letters indicating that it is designed for application at moderate temperatures (MX) and that this red has a pronounced bluish shade (5B). These suffixed letters are most important in dye nomenclature. Moving on from these high reactivity dichloro-*s*-triazines,

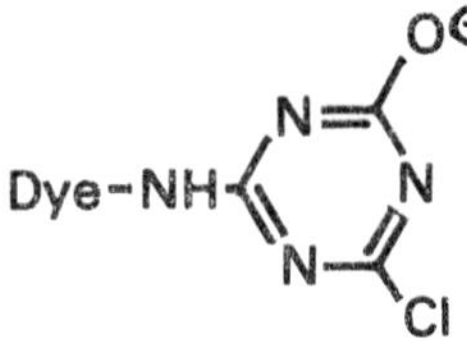

(VI)

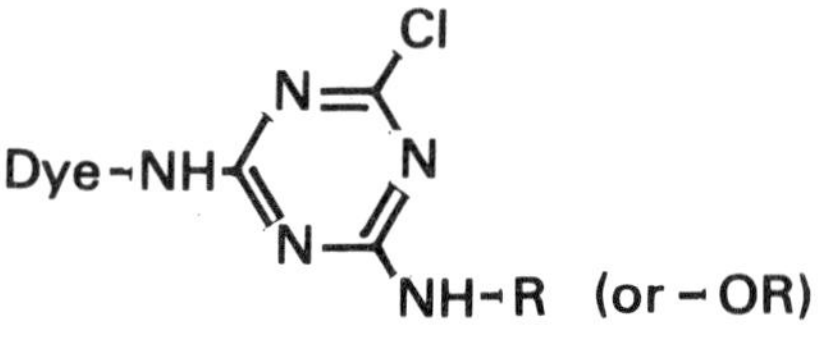

(VII)

replacement of a further chlorine atom by an electron-donating group, e.g. amino, substituted amino or alkoxy (VII), diminishes the level of reactivity of the remaining chlorine atom. Dyes of this type are to be found in the PROCION H, PROCION P, PROCION H-E and CIBACRON ranges of reactive dyes. They can be applied under the same sort of exhaust dyeing conditions as those outlined above for the dichloro-*s*-triazinyl dyes but with the difference that the sequence of steps is carried out at about 85°C and usually slightly more alkaline conditions are required for satisfactory fixation (pH 11 vs. pH 10.5 for the dichlorotriazines). There are good technical reasons for the development of this type of dye which lie rather outside the scope of the present discourse. Suffice it to say that operation at a higher temperature allows a bigger dye molecule to be employed, the elevated temperature allowing the requisite mobility of the dye within the fibre to be maintained. Ideally, in the PROCION H-E type two reactive groups are incorporated in order to get a greater proportion of the dye to fix rather than hydrolyse. As an alternative to raising the temperature, the reaction time can be prolonged to allow fixation to occur at room temperature. This option, requiring perhaps four days for full fixation to occur is not one which appeals to the textile dyer, to whom time is money. For biochemical uses, however, it presents a perfectly acceptable option and has a possible advantage in that no thought need be given to possible complications arising through subsequent hydrolysis or other reaction of the third chlorine atom which remains on the attached dye after a dichlorotriazine dye has been used (Dawson *et al.* 1962).

One facet of triazine chemistry which might have presented some attraction to biochemists is the fact that by forming a quaternary ammonium salt between a monochlorotriazine dye and a suitable tertiary amine (VIII), the reactivity of this species can be boosted to the level where dye–substrate reaction occurs at

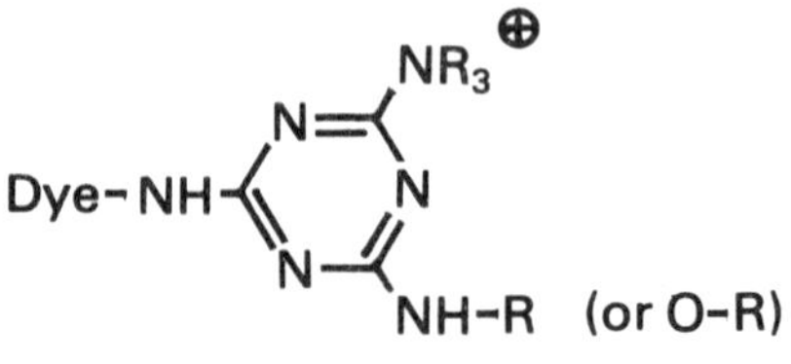

(VIII)

room temperature (Dawson 1964). Formation of these quaternary salts occurs so readily that there is no need to preform the product and the effect can be achieved merely by adding the requisite tertiary amine to the dye solution.

Viewing the reactive dye as being composed of two parts, namely the chromogen and the reactive system, these two fragments have different purposes. The reactive system serves to anchor the dye to the cloth and thus frees the chromogen from any requirement other than the need to contain sufficient sulphonic acid groups to give adequate water solubility and also an amino group to which the reactive system can be attached. This enables any class of dye chromogen to be used. In practice the azo, anthraquinone, phthalocyanine and triphendioxazine types predominate. This apparently bewildering range of structural types must make it difficult to correlate biochemical effects with dye structure. Two trends seem to have emerged. First, the larger dye structures found in the PROCION H-E range have displayed particular utility. Possibly they owe their efficacy to parts of the aromatic system being spaced further away from the substrate backbone than is the case with simpler types. More satisfying is the position with the closely interrelated PROCION Blue H-B and Cibacron Blue F-3GA both of which are anthraquinone derivatives whose attached form corresponds to (IX). Structural studies here have shown a considerable similarity between these molecules and the adenosine unit of the NAD co-factor (X). In affinity chromatography applications the protein simply mistakes the anthraquinone moiety for the relevant co-factor resulting in selective protein adsorption onto the dye–substrate matrix.

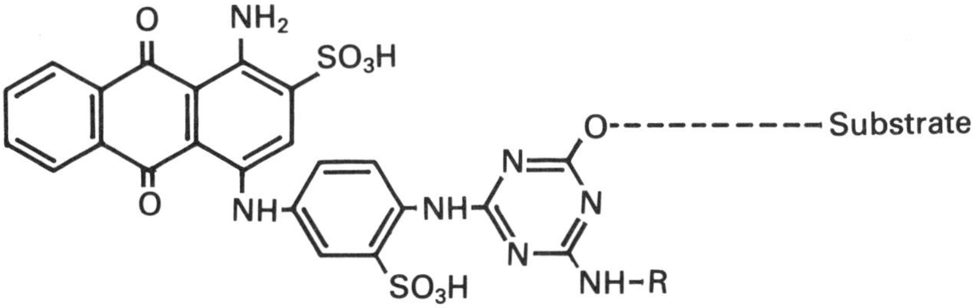

(IX)

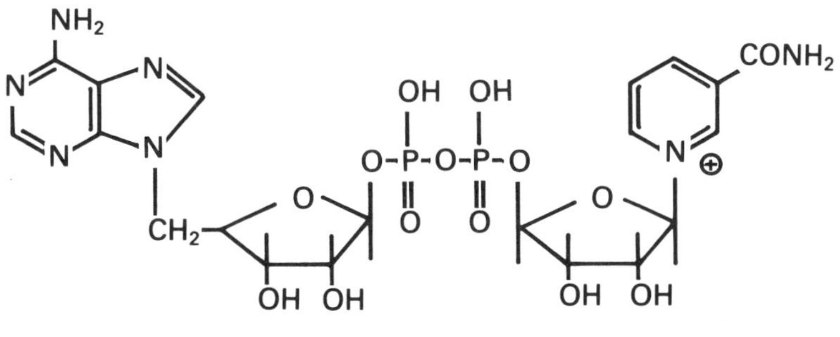

(X)

10.2 THE BIOLOGICAL APPLICATIONS OF TRIAZINE DYES
(C. R. Lowe)

The selective interaction of proteins with chlorotriazine dyes was first noted in 1968 when it was shown that the anthraquinone dye Cibacron blue F3G-A could bind to certain enzymes (Kopperschläger *et al.* 1968; Haeckel *et al.* 1968). Interest centred on this particular dye because a dextran conjugate of Cibacron blue F3G-A had been used as a void volume marker for gel filtration for many years and it was observed that yeast pyruvate kinase behaved anomalously when co-chromatographed with blue dextran on a column of Sephadex G-200. Subsequent investigations revealed that the appearance of the enzyme in the exclusion volume could be attributed to its interaction with the chromophore rather than the dextran carrier itself (Kopperschläger *et al.* 1971). Within a short time, a similar procedure involving the combination of blue dextran and gel filtration, was used to purify a wide range of proteins, and techniques were soon developed to covalently bond blue dextran directly to a suitable solid phase, such as agarose. More recently, the blue chromophore, Cibacron blue F3G-A, has been immobilised to a variety of supports, including Sephadex, polyacrylamide, starch and cellulose and used for the purification of a host of diverse proteins by affinity chromatography (Ryan & Vestling 1974). Indeed, interest in commercial textile dyes as ligands for affinity chromatography has risen dramatically in the last few years and a cumulative total of well in excess of 1000 papers now purport to purify 500 or more different proteins. Thus, examples of every class of enzyme together with a number of seemingly unrelated non-enzymic proteins have been shown to interact selectively with chlorotriazine dyes (Lowe 1984a).

It is not immediately obvious why synthetic printing and textile dyes are able to interact with complementary proteins with such a degree of specificity. It is now generally agreed, however, that dyes mimic the overall size, shape, aromaticity and anionicity of natually occurring biological heterocycles such as coenzymes, nucleotides and polynucleotides. Thus, evidence from direct kinetic evaluation of dye–protein interactions, electrophoresis, affinity labelling, difference spectroscopy, circular dichroism, NMR spectroscopy and X-ray crystallography all point to dye binding at the active site of proteins. For example, a substantial body of data has been collected on the ability of reactive dyes to act as reversible inhibitors of enzyme catalysis. For many enzymes, particularly the dehydrogenases and kinases, the inhibition was competitive with respect to the nucleotide substrates. In fact, the anthraquinone dye, Cibacron blue F3G-A, has been shown to be competitive with the binding of several nucleotides (GTP, ATP, IMP), cyclic nucleotides (cAMP, cGMP), polynucleotides (DNA), pyridine nucleotide coenzymes (NAD^+, $NADP^+$), flavins (FMN, FAD), folates, Coenzyme A esters, pyridoxal coenzymes and some substrates such as 3-phosphoglycerate. In most cases, the inhibition constants (K_1) were in the low micromolar range and therefore often considerably less than the K_m value for the natural substrates.

The perturbation of the visible adsorption bands of dye chromophores on binding to proteins provides an additional experimental handle to monitor dye–protein interactions. As a general rule, the absorption spectrum of Cibacron blue

F3G-A experiences a pronounced red shift of 50–70 nm from the absorbance maximum of 610–620 nm in the presence of some proteins. Difference spectra recorded in the presence or absence of salts, chaotropes, hydrophobic agents or specific substrates, nucleotides or coenzymes permit further assessment of the dye–protein interaction. For example, the interaction of Cibacron blue F3G-A with rabbit muscle lactate dehydrogenase induces a 50 nm red shift in the adsorbance maximum of the dye which can subsequently be removed on addition of NAD^+ or NADH. These observations confirm that Cibacron blue F3G-A binds to the coenzyme-binding site of lactate dehydrogenase. Similarly, both NAD^+ and NADH reduced the magnitude of the induced circular dichroism spectrum and yielded information on changes in molcular configuration as a result of dye binding and the detailed topography of the binding site (Edwards & Woody 1979; Towell & Woody 1980). Studies with several other proteins revealed that the principal chromophoric group of Cibacron blue F3G-A, namely the anthraquinone moiety, is intimately involved in all protein binding, although the detailed architecture and microenvironment varies from protein to protein.

Several studies have attempted to define the structural elements required for dye binding by utilising either analogues or fragments of Cibacron blue F3G-A (Lowe 1984a). The dye may be divided into four distinct structural elements: (a) the substituted antraquinone moiety, (b) the bridging diaminobenzene sulphonate, (c) the reactive chlorotriazine ring and (d) the terminal aminobenzene sulphonate ring. Kinetic inhibition studies with several Cibacron blue F3G-A analogues and a number of enzymes revealed that the anthraquinonyl end of the chromphore, i.e. rings (a) and (b) were crucial for strong potency as an inhibitor. The anthraquinone ring alone (a) was a poor inhibitor of several oxido-reductases and kinases whilst a fragment comprising rings (a) and (b), (1-amino-4,4′-aminophenylamino)-anthraquinone-2′,3′-disulphonic acid, was only slightly less inhibitory than the parent dye, Cibacron blue F3G-A, itself. The importance of the anthraquinone moiety as the key structural feature of the dye was confirmed by studies on the binding of proteins to four different anthraquinone dyes (Beissner & Rudolph 1978; Land & Byfield 1979) and to Cibacron blue F3G-A immobilised through different functional groups (Beissner & Rudolph 1978).

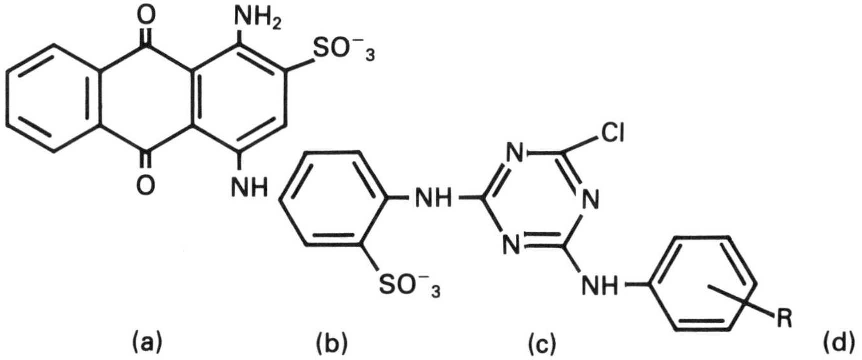

Perhaps the most conclusive evidence that reactive dyes mimic, at least partially, the naturally occurring biological ligands such as NAD^+, $NADP^+$, nucleotide mono-, di- and triphosphates, flavins, coenzyme A and folates comes from affinity labelling studies. It is not unreasonable to expect that the free dyes should act as irreversible active site-directed affinity labels and therefore yield valuable information on dye–protein binding. Initial studies showed that a number of reactive dichlorotriazinyl dyes specifically and irreversibly inactivated pig heart lactate dehydrogenase, yeast glucose-6-phosphate dehydrogenase and yeast hexokinase at sites competitive with the coenzymes NAD^+, $NADP^+$ and ATP respectively (Clonis & Lowe 1980). Dichlorotriazinyl dyes inactivate pig heart lactate dehydrogenase in a time-dependent fashion, display a hyperbolic dependence on dye concentration and show protection with specific nucleotides. Affinities are highest for blue-green dyes and lowest for yellow-orange dyes. These observations suggest that all the dyes bind to the NAD^+-binding site but with markedly unequal affinities. Similar results were obtained with yeast hexokinase, although of a number of dyes tested, the copper phthalocyanine dye, Procion green H-4G was apparently unique (Clonis *et al.* 1981). The dye inactivated yeast hexokinase in a fashion competitive with the adenine nucleotides ATP and ADP and the sugar substrates D-glucose, D-mannose and D-fructose. Quantitatively inactivated enzyme contained one mole dye per subunit of molecular weight 51000. In addition, the activation of the enzyme by Procion green H-4G was inhibited in the presence of Mg^{2+} ions. It is believed that Mg^{2+} ions increase the affinity of the dye for the enzyme about five-fold and promote the binding of the Mg^{2+}-Procion green H-4G complex across the inter-domain cleft between the nucleotide and sugar binding site.

More detailed analysis showed that the inactivation of horse liver alcohol dehydrogenase by Procion blue MX-R was competitively inhibited by NAD^+ and NADH and resulted in the incorporation of one mole dye per mole subunit of molecular weight 40000 (Small *et al.* 1982). Removal of excess dye, chymotryptic digestion, resolution of the peptides by reverse phase, high performance liquid chromatography and sequencing of the purified blue peptide yielded a pentapeptide with the affinity label, Procion blue MX-R covalently attached to the side chain thiol of cysteine-174. This important residue lies at the bottom of the active site cleft between the catalytic and nucleotide domains and normally provides one ligand, the thiol side chain, to chelate the catalytic zinc atom. This observation confirms that Cibacron blue F3G-A is bound to horse liver alcohol dehydrogenase in a manner reminiscent of the coenzyme with the anthraquinone moiety lying in the apolar adenosine binding pocket and the chlorotriazine innermost. This conclusion concurs with X-ray diffraction studies on the binding of Cibacron blue F3G-A to crystals of liver alcohol dehydrogenase at 0.37nm resolution (Biellmann *et al.* 1979). The anthraquinone ring (a) binds in a broad pocket comprising the adenine binding site at one end, whilst the bridging ring (b) binds in such a position that its sulphonate group interacts with the guanidinium side chain of Arg-271. The chlorotriazine ring (c) binds close to where the pyrophosphate moiety of the coenzyme binds with the chlorine close to the binding site for the nicotinamide ribose. The terminal ring (d) is bound in a cleft between the catalytic and nucleotide domains with a likely interaction of the anionic sulphonate with the side chain of Arg-369. It appears that the binding of Cibacron blue F3G-A to horse liver alcohol dehydrogenase parallels the binding of the ADP moiety but differs significantly from the nicotinamide end.

The mechanistic and labelling studies described above provide irrefutable evidence that chlorotriazine dyes such as Cibacron blue F3G-A bind to the coenzyme binding site of certain oxidoreductases and strongly suggests that similar interactions might occur for other naturally occurring biological ligands. Not surprisingly, therefore, the bio-mimetic textile dyes have found widespread application in protein purification (Lowe 1984a; Lowe *et al.* 1981). Their ready availability in large quantities and with a variety of different chromophores, their cheapness, ease of immobilisation, broad specificity and stability to chemical or biological attack has contributed to their general acceptance as ligands of choice for the large-scale purification of proteins. For example, immobilised Cibacron blue F3G-A has now been used for the purification of oxido-reductases, kinases, coenzyme A-dependent enzymes, restriction endonucleotides, synthetases, lyases, phosphatases, decarboxylases and a variety of blood proteins (Lowe 1984b). Thus, albumin has been purified from plasma, Cohn Fraction IV and red blood cell washings by affinity chromatography on Cibacron blue F3G-A Sepharose CL-6B. Albumin in excess of 95% purity and in yields greater than 90% was obtained. An automated pilot scale system exploiting a 16-litre column binding 50 g albumin/litre of blue Sepharose has now been established.

More recent developments have exploited reactive dyes in a number of novel manifestations of the affinity technique. For example, reactive dyes have been covalently attached to silanised microparticulate silica and exploited in high performance liquid affinity chromatography (HPLAC). This technique combines the speed of high performance liquid chromatography with the biological specificity of affinity chromatography and permits the resolution of proteins, enzymes and isoenzymes within minutes (Lowe 1984a; Lowe 1984b). More recently, the reactive anthraquinone dye, Procion blue MX-R, has been covalently attached to preparative grade silica (40–63 μm) and exploited for the rapid large-scale purification of rabbit muscle lactate dehydrogenase by preparative HPLAC (Small *et al.* 1983). Application of a partially purified rabbit muscle preparation yielded essentially homogeneous lactate dehydrogenase in 80% overall yield in less than 15 min. The technique displays considerable potential for the rapid large-scale isolation and resolution of proteins.

The volume capacity of most solid phase technologies is often a serious limitation to their eventual scale-up into commercially viable processes, although homogeneous technologies can dramatically improve this situation. For example, the techniques of affinity precipitation in aqueous two-phase systems and affinity precipitation with bifunctional affinity ligands both display promise for improving the volume capacity of the processes (Lowe 1984b). In the future, one would expect to see the synthesis of novel dye structures which are more precise analogues of the natural biological ligands and thereby are more selective for the purification of individual proteins. It is anticipated, therefore, that reactive dyes will develop further as popular and colourful additions to the armoury of the protein purification technologist.

REFERENCES

Beech, W. F. (1970) *Fibre Reactive Dyes*, London, Logos Press.

Biellmann, J.-F., Samama, J.-P., Bränden, C. I. & Eklund, H. (1979) *Eur. J. Biochem.* **102**, 107–110.

Beissner, R. S. & Rudolph, F. B. (1978) *Archs Biochem Biophys.* **189**, 76–80.

Clonis, Y. D., Goldfinch, M. J. & Lowe, C. R. (1981) *Biochem. J.* **197**, 203–211.

Clonis, Y. D. & Lowe, C. R. (1980) *Biochem. J.* **191**, 247–251.

Dawson, T. L. (1964) *J. Soc. Dyers and Colorists* **80**, 134; also UK Patent 937 182 (ICI).

Dawson, T. L., Fern, A. S. & Preston, C. (1962) *J. Soc. Dyers and Colorists* **78**, 274.

Edwards, R. A. & Woody, R. W. (1979) *Biochemistry* **19**, 5197–5204.

Haeckel, R., Hess, B., Lauterborn, W. & Wuster, K.-H. (1968) *Hopper-Seyler's z. Physiol. Chem.* **239**, 699–714.

Horrobin, S. (1963) *J. Chem. Soc.* 4130 also UK patents 838,337, and 842,193.

ICI, UK Patents 797,496, 798,121 and many subsequent patents.

Ingamells, W., Sumner, H. H. & Williams, G. (1960) *J. Soc. Dyers and Colorists* **76**, 210.

Kopperschläger, G., Freyer, R., Diezel, W. & Hofmann, E. (1968) *FEBS Lett.* **1**, 137–141.

Kopperschläger, G., Diezel, W., Freyer, R., Liebe, S. & Hofmann, E. (1971) *Eur. J. Biochem.* **22**, 40–45.

Land, M. & Byfield, P. G. H. (1979) *Int. J. Biolog. Macromol.* **1**, 223–226.

Lowe, C. R. (1984a). In: Wiseman, A. (ed.), *Topics in Enzyme and Fermentation Biotechnology*, vol. 9. Chichester, Ellis Horwood, pp. 78–161.

Lowe, C. R. (1984b) *J. Biotechnol.* **1**, 3–12.

Lowe, C. R., Small, D. A. P. & Atkinson, A. (1981) *Int. J. Biochem.* **13**, 33–40.

Ryan, L. D. & Vestling, C. S. (1974) *Archs. Biochem. Biophys.* **160**, 279–284.;

Small, D. A. P., Lowe, C. R., Atkinson, A. & Bruton, C. J. (1982) *Eur. J. Biochem.* **128**, 119–123.

Small, D. A. P., Atkinson, A. & Lowe, C. R. (1983) *J. Chromatogr.* **266**, 151–156.

Towell, J. F. & Woody, R. W. (1980) *Biochemistry* **19**, 4231–4237.

Venkataraman, K. (ed.) (1972) *The Chemistry of Synthetic Dyes*, vol. VI. London, Academic Press.

The main source of information relating to dyestuffs is the patent literature, but it is confusing to the non-expert. The books by Beech and Venkataraman are recommended for a fuller discussion of the subject of reactive dyes.

11

Isolation and recovery of enzymes

P. Fullbrook and D. Parker
Imperial Biotechnology Limited,
263–269 City Road, London EC1V 1JX, UK

11.1 INTRODUCTION

So far, this book has focussed on several approaches and developing techniques for the isolation and recovery of microbial products. Several chapters have already described the application of specific techniques – in some detail – for the isolation of enzymes and other proteins. As no two enzymes will be isolated in exactly the same way, the task of reviewing all the methods used for their isolation and recovery is superfluous. Instead, this chapter concentrates on the discovery and isolation of enzymes – principally from microbial sources – which have defined, and practical use applications.

The chapter falls into two logical parts: (1) a statement of the principles involved, plus an illustrative review of progress made from the beginnings of biochemistry in the nineteenth century up to and including the 1970s and (2) a look at more recent developments – with specific reference to the newer isolation methods and the application of genetic engineering to enzymology.

11.2 METHODS FOR COMMERCIAL ENZYME PRODUCTION

The specific methods adopted for the isolation and recovery of enzymes – especially those enzymes intended for commerical application – will depend upon a variety of factors. Usually, several methods will be used in logical sequence. These will generally be chosen from the sort of methods already described elsewhere in this book.

The most important factor in choice of purification regime, however, is consideration of the final use, or application to which the enzyme product is to be fitted. Enzymes are generally of use only for their catalytic or identification properties, i.e. their chemical functionality. Their protein nature, as far as their use is concerned, is almost incidental. Because of their unique specificity, their in-use purity is also of secondary importance. Essentially the end use of the enzyme will not only determine, the degree of purification required, but it will also have an important bearing on the cost of the enzyme in use. This cost-effectiveness will set the frame for the feasibility of enzyme production, in relation to production and purification costs. In terms of isolation and recovery of enzymes therefore, the most important factor is enzyme product specification. This has to mirror, and carefully match its use application.

By way of illustration, it is important to realise that an enzyme to be used clinically, or for an analytical or diagnostic application, has to have a much tighter, better defined specification in terms of its purity and specificity, than an enzyme which is to be used as a transformation catalyst – for example, in the production of a bulk chemical or intermediate form of cheap agricultural feedstock. As is shown later, most commercially used enzymes today fall into this latter category, and hence require minimal purification. The emphasis then is on cheap isolation and efficient, effective recovery methods – essentially a concentration and formulation problem.

Table 11.1 compares the composition of many typical industrial enzymes, with that of an enzyme used in clinical diagnosis. The higher specific activity and requirement for complete lack of competing side activity enzymes for the diagnostic enzyme is also evident. The relatively high sugar/inorganic salts levels in the preparations are indicative of these substances being used as stabilisers in both product types. In the case of the particular diagnostic enzyme illustrated, the sugar is sucrose and is in effect acting both as a stabiliser and diluent.

Table 11.1 – Comparison of the composition of commercial enzyme preparations.

	Range of contents (% dry weight)	
Enzyme type	Industrial	Diagnostic
Proteins + amino acids	10–15	35
Active enzyme protein	2–5	33
Complex carbohydrates	5–12	0
Sugars	2–40	55
Inorganic salts	3–30	5
Preservatives	0–0.3	0
Secondary activities	Low	Absent

Data is based partly on information from Godfrey & Reichelts, *Industrial Enzymology*, 1983, p. 4, (Macmillan–Nature Press, UK).

As might be expected, the fermentation costs for the production of, for example, amyloglucosidase (EC: 3.2.1.3) – an industrial enzyme used for the production of glucose from liquefied starch – are broadly comparable with those costs for fermentation of a bacterial diagnostic enzyme, e.g. glutamic dehydrogenase, (EC: 1.4.1.3), used for the clinical determination of blood urea nitrogen. In contrast, however, the downstream processing costs of recovery, and purification to a higher specification, are much greater for the latter enzyme. The same would be essentially true for the production of high value-added products, produced by fermentation – such as recombinant γ-interferon produced by *Escherichia coli*, or the various restriction endonuclease enzymes used in genetic recombination technology. These would, however, probably have higher medium ingredient costs.

Although, initially enzymes were recovered from animal and plant sources, most commercially used enzymes today, are of microbial origin. This is because of the remarkable potential of microorganisms to be produced in virtually unlimited quantities by fermentation processes, under carefully controlled uniform conditions, fom cheap, and plentiful raw materials (agricultural feedstocks and commodity chemicals).

The main factors affecting the isolation and recovery methods for microbial enzymes are outlined in Table 11.2. They comprise the characteristics of the microorganism expressing the particular enzyme activity required, and the characteristics of the enzyme itself. Note that the final choice of isolation methods used will be influenced as much by the end use of the recovered enzyme and economic factors associated with the downstream purification as by these biochemical and technical factors.

Quite obviously, the higher the concentration of the required enzyme that is expressed during the fermentation stage, the easier it will be to meet a given product specification – although there are exceptions to this in practice. Fig. 11.1 illustrates the point for a variety of reference biological products and enzymes. It shows the roughly regular (log/log) basis for the relationship between concentration of metabolite occurring in the cultivation state and the eventual product selling price across the spectrum from commodity service products to high added-value clinical products. Interestingly, it also represents the timescale for the development of separation biotechnology. The left-hand side of the graph represents the mature phase of development where separation or purification is effected by mechanical, civil or chemical engineering unit operations and includes conventional distillation, precipitation or simple extraction. The centre portion represents the current growth phase employing state-of-the-art biotechnology using techniques such as crystallisation, ultrafiltration and non-specific chromatographic separations. The right-hand side illustrates the embryonic phase of development dependent on new technology: either techniques of the new biotechnology such as recombinant or affinity chromatographic technologies or multi-step advanced conventional downstream processing. Enzyme isolation and purification spans these latter categories with industrially used products tending towards the mature phase whilst newer products for research, diagnostic or therapeutic uses are included in the higher value-added products towards the right-hand side.

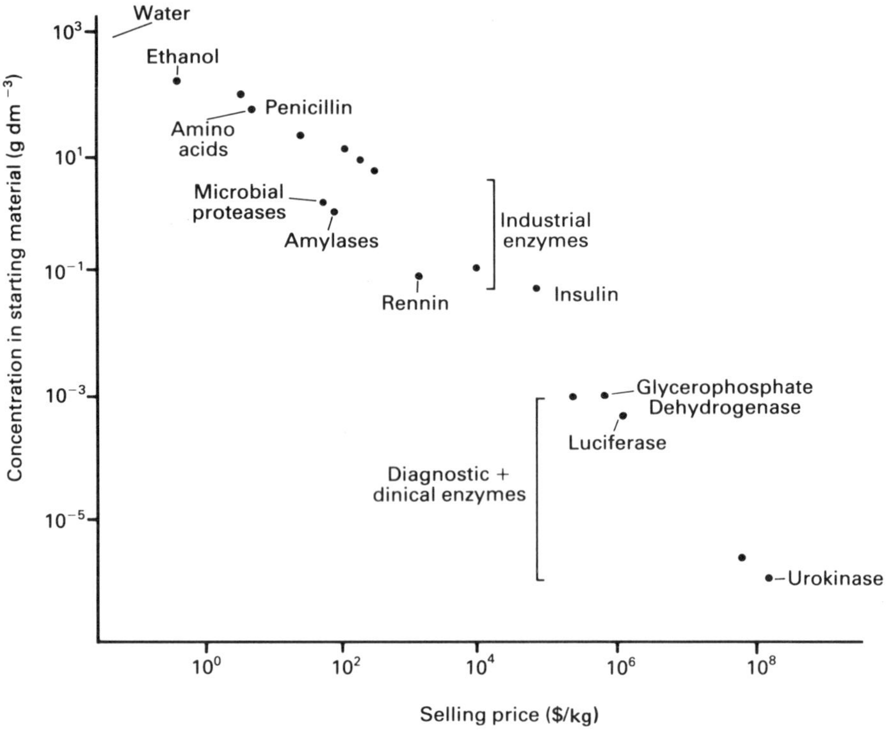

Fig. 11.1 – Concentration in starting material of various substances and their selling price. (Data from J. L. Dwyer, *Biotechnology*, November 1984).

The advent of genetic engineering has afforded the possibility for programmed over-production of specific enzyme activities. Over-production of enzymes is not however confined to recombinant products, as Fig. 11.2 illustrates. Careful selection techniques of conventionally produced mutants can give similar high yields of enzymes, making isolation and recovery a much easier task. The illustration shows an SDS-PAGE run on a bacterial extract containing glutamate dehydrogenase – an enzyme we are currently producing. Tracks show 2 concentration levels comprising (from right to left) MW marker proteins, crude culture filtrate, and pooled, desorbed column eluate – the purified product – after dye–ligand chromatographic separation.

Table 11.2 – Factors affecting the isolation and recovery methods of microbial enzymes.

1.	Characteristics of microorganism expressing enzyme	
	Bacterial Fungal Recombinant	Pathogenicity/toxicity
	Growth	nutritional requirements aerobic/anaerobic constitutive/inducible product inhibited
2.	Characteristics of expressed enzyme(s)	
	Intracellular Extracellular Microsomal/particle bound	
	Level of expression	
	Chemical nature of protein	activation/pro-enzyme monomer/polymer stability competing contaminants

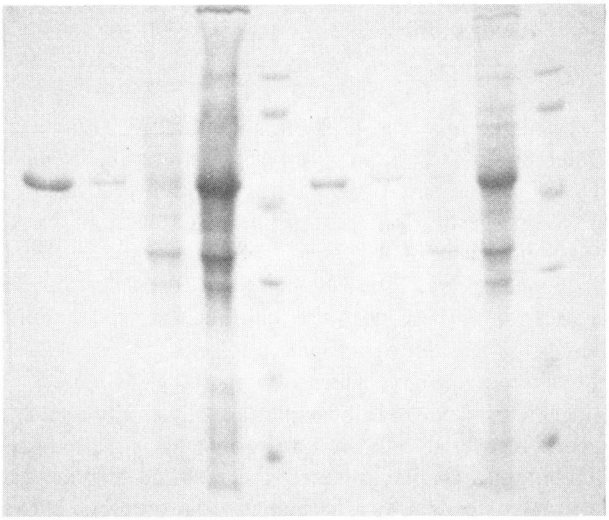

Fig. 11.2.

Although most industrial enzymes are produced similarly from high yielding, conventionally mutated strains – e.g. proteases and amylases – some animal and plant enzymes can now be expressed from recombinant microorganisms – e.g. calf rennet in *E. coli* and cereal β-amylase in *Bacillus subtilis.*

The main strategies employed in designing operating procedures for the isolation and recovery of microbially produced enzymes, are summarised in Table 11.3. As losses occur at each and every successive unit operation, the necessity to minimise the number of operational steps, is obvious. Exploiting the unique properties of the enzyme under isolation, is one elegant way of achieving this. Simplicity of operational methodology is also an obvious way of increasing the reliability of batch processing by improving the reproducibility and repeatability of the steps involved. The actual techniques selected for isolation and recovery, will be different for each and every individual enzyme, or indeed enzyme application. As it is not yet possible to patent an enzyme as such, enzyme production methods are either heavily protected by wide patents, or else remain the confidential and proprietary information of the producer. We will not therefore go into detail, but just comment on the general principles involved.

Table 11.3 – Useful guideline strategies in designing operating procedures for the isolation and recovery of enzymes.

Keep it simple
Keep it cheap
Minimum number of unit operations
Maximal exploitation of the unique properties of the enzyme activity being purified

Until recently, the majority of industrial enzymes were isolated as extracellular products from simple fermentations of *Bacillus* or *Aspergillus* species. Isolation was effected by removal of producing cells by mechanical means of centrifugation or drum filtration, often assisted by the use of flocculating agents and specific pH/ionic strength conditions. The liquors obtained were then concentrated, often by means of ultrafiltration, and standardised to the specification activity with the addition of added stabilisers and preservatives – all conventional unit operations.

Bacillus coagulans glucose isomerase, (EC: 5.3.1.5) was one of the first intracellular enzymes to be exploited industrially on any significant scale. In one process, recovered cells were mechanically harvested and broken to release the cell contents which contained the required enzyme activity. They were then heat-treated to destroy a contaminant proteolytic activity which would otherwise inactivate the enzyme protein. The surviving glucose isomerase activity was then conveniently absorbed back onto the cell debris, and fixed by crosslinking with a bifunctional chemical reagent such as glutaraldehyde to produce a recoverable, reusable immobilised enzyme. *Escherichia coli* penicillin amidase (EC: 3,5,1,11) was first used as a cell-bound enzyme catalyst, and is another industrially significant enzyme product. The newer conventional intracellular enzymes are isolated in processes similar to the scheme shown in Table 11.4. The objective is to free the intracellular enzyme of any solid material and to produce semi-pure, soluble concentrates which can be formulated to give a standard product.

Table 11.4 – General procedure for isolation and purification of intracellular enzymes.

Recover cells from fermentation broth
Break cells by homogenisation/disruption
Wash
Remove debris and particulate matter
Concentrate isolated liquor
Fractionate by specific absorption/desorption and/or precipitation techniques
Standardise and formulate

As the theme of this book is the discovery and isolation of microbial products, this section is concluded by tracing the development of enzymology in the practical use of isolated enzymes. Fig. 11.3 shows the number of enzymes recognised and characterised since 1853 when the first reported enzyme was isolated as an aqueous extract of barley malt by Payen and Persoz. It was precipitated by ethanol as a heat labile substance and had the ability of hydrolysing starch to sugars. (In reality is was a mixture of several amylases, which were termed Diastase).

Even though today we can recognise several thousand different enzymes, and understand the exact chemical structure and mechanism of quite a few of them, there are only a few dozen which are exploited in a large-scale industry, as is illustrated in Figs. 11.4 and 11.5. These show the production volumes of the five most important commercially used enzymes. All these are produced as rather crude products for the processing of food and domestic materials, and require little purification after isolation from the fermenter.

11.3 PROTEIN ENGINEERING

11.3.1 General considerations

In recent years, recombinant DNA technology has advanced to the point where we can in theory clone the gene for any protein found in nature. With recent improvements in the chemical synthesis of DNA we can also perform extensive genetic modifications of cloned genes. Thus there is the potential to produce

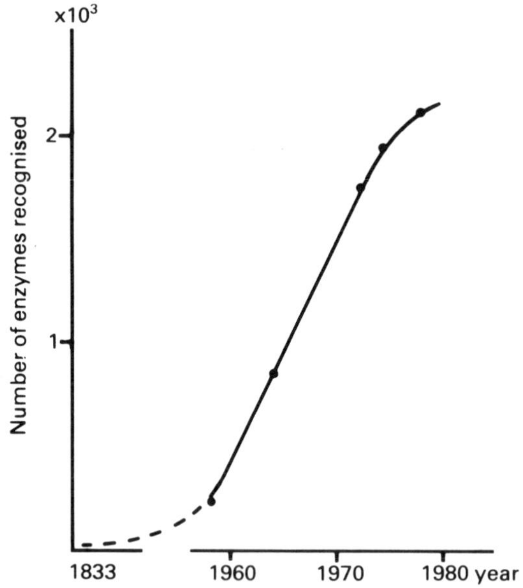

Fig. 11.3 – The discovery of enzymes. (Information from M. Dixon and E. C. Webb, *Enzymes*, Longmans, London, 1979).

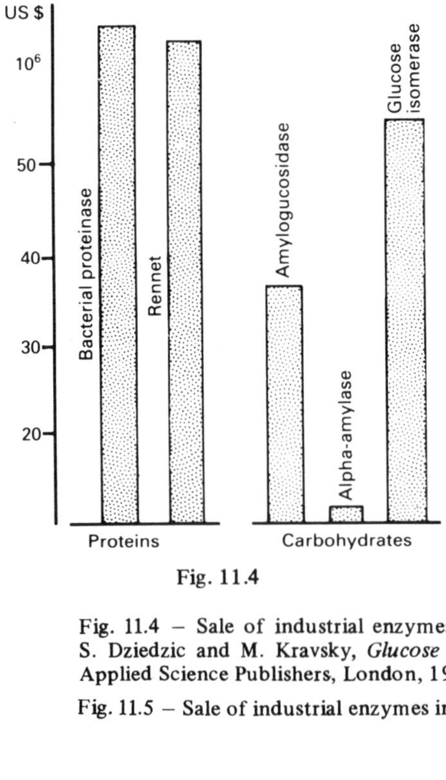

Fig. 11.4

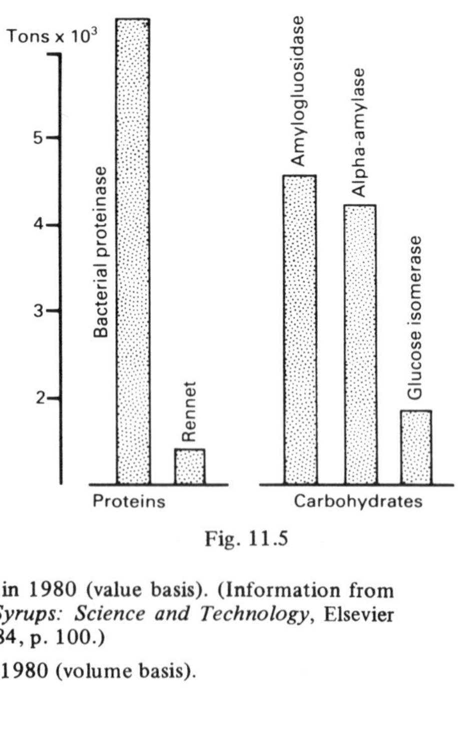

Fig. 11.5

Fig. 11.4 – Sale of industrial enzymes in 1980 (value basis). (Information from S. Dziedzic and M. Kravsky, *Glucose Syrups: Science and Technology*, Elsevier Applied Science Publishers, London, 1984, p. 100.)

Fig. 11.5 – Sale of industrial enzymes in 1980 (volume basis).

recombinant and genetically altered proteins in microorganisms in high yield and at realistic cost. This section will briefly review the prospects for protein engineering in industrial enzymology before examining the fundamental problems in downstream processing which have prevented the widespread exploitation of the technology for the production of eukaryotic proteins in *E. coli.*

Protein engineering is the logical successor to random screening of microorganisms and mutation/selection programmes for the discovery and improvement of biological catalysts. It is the planned, deliberate alteration of DNA sequences so as to change the primary amino acid sequence and thus endow the resultant protein with new properties. Some applications of this new technology are given below.

Protein engineering will enable the production of enzymes with:

altered kinetic properties, e.g. K_m, V_{max}, K_{cat};
new substrate and reaction specificity;
altered cofactor requirements, e.g. NADH instead of NADPH;
changed pH optima;
thermostability and altered temperature optima;
resistance to oxidation;
stability in non-aqueous solvents;
altered quaternary structure;
removal of allosteric regulation.

Although some advances can be made by educated guesswork, to engineer proteins in a reasonably predictable fashion it is necessary to utilise a considerable amount of data and physical resources. A cloned gene for the particular enzyme must of course be available. The amino acid sequence may be determined by inference from the nucleotide sequence of the gene. The three-dimensional structure of the protein must be solved using X-ray crystallographic techniques and a computer graphics display facility may be required to model the protein structure and to design protein structural changes. Certainly one must have knowledge of the solution chemistry of the enzyme at the molecular level to be able to design alterations to tailor the protein to a new or improved process. All these technical areas have seen rapid developments in recent years. At present, the slowest steps are growing suitable protein crystals for high resolution X-ray analysis and solving the crystal structure. However, computer modelling methods are advancing rapidly so it may not be necessary to solve the protein structure but to estimate it using computation and comparisons with similar proteins of known structure. For example, Blundell *et al.* (1983) have modelled the structure of the aspartic proteinase renin using interactive computer graphics techniques. The model was based upon the primary amino acid sequence of renin and the known three-dimensional structure of another aspartyl proteinase, endothia pepsin, with which it shared sequence homology. It is commonplace to discover that enzymes catalysing the same or similar reactions, even those from diverse organisms such as bacteria and mammals, have extremely similar sequences, structures and enzyme mechanisms. Therefore it will not be necessary to solve crystal structures for every enzyme under consideration.

Although it will become possible to design new proteins from first principles, it is doubtful whether the intensive effort will be worthwhile since nature has provided us with a diversity of protein structural domains through evolution and has in effect performed the experiments for us. Therefore domain structures from different proteins could be fused together to form the basic structures of novel proteins.

To date there have only been a handful of published accounts of successful protein engineering experiments. Particularly notable examples are the work of Wilkinson *et al.* (1984) who achieved substantial improvements to the kinetic constants of the enzyme tyrosyl tRNA synthetase, and the modification of a methionine residue in α_1-antitrypsin by Rosenberg *et al.* (1984) to confer oxidation resistance to the molecule. These examples have established the utility of the technology.

To enable us to design and produce highly stable supercatalysts it will be necessary to undertake further fundamental work to increase our knowledge of enzyme mechanisms and the forces which stabilise the folded structure of proteins, particularly the thermodynamic aspects which enable a catalyst to stabilise transiently the transition state of a substrate. One anticipates that the application of protein engineering will allow much new information to be accumulated on these subjects and will bring immense benefit to both industry and academia.

11.3.2 Exploitation of recombinant DNA technology

Although the introduction of new products produced by protein engineering is still a few years away, there has been considerable activity and publicity associated with the more conventional applications of recombinant DNA technology. This has centred on the potential to produce rare and valuable proteins on a large scale, e.g. interferon. However, few products have yet been launched in the marketplace. Among the reasons for this are the gross difficulties which have arisen during scale up. The major problem that has inhibited the large-scale production of eukaryotic proteins in *E. coli* is one of product expression. There were initial difficulties in persuading the bacterial synthetic machinery to produce protein from the foreign DNA sequences. The problems have generally been overcome, either by new gene constructions or by the tactic of fusing the DNA encoding the desired product to the protein-coding region of an *E. coli* gene which contains a controllable promoter and is known to produce high levels of protein. In many cases a small portion of structural gene suffices and the fusion provides control of expression and a 'slip road' into the desired coding region. The resultant product, known as a fusion protein, will therefore contain some amino acid residues at the N-terminus which emanate from the bacterial protein. These residues may need to be removed before such a product can be used as a therapeutic agent. A useful review of the subject has been given by Harris (1983).

Unfortunately, the desire to produce large quantities of foreign protein in bacteria has led to a further severe problem. When the cloned, plasmid-born genes are activated during a fermentation there is a massive synthesis of product protein. This protein does not, by and large, remain soluble within the cell, but aggregates into large particles known as inclusion bodies. During centrifugation following cell rupture these inclusion bodies are sedimented along with the cellular debris. Since the waste supernatant fraction following centrifugation contains essentially all of the soluble proteins this might appear to be an advantage regarding product purification, as indeed it would be if the product could be solubilised under moderate conditions. However, this is rarely the case as the form of the product within the inclusion bodies is that of insoluble denatured protein. The case of solubilisation of the product is somewhat dependent upon the protein in question but in general terms a strong protein solvent is required such as concentrated urea, guanidine hydrochloride, alkali, acid or ionic detergents such as sodium dodecyl sulphate. Thus the product is solubilised from a denatured state and maintained in a denatured form by the solvent. There is an additional penalty in that water-insoluble membrane bound bacterial proteins (for example) are also solubilised to a greater or lesser extent and contaminate the preparation.

The downstream processing of such material thus poses a great many new problems since the established techniques in protein purification have been designed for operation with native folded proteins using water as a solvent. The major purification advances of the last two decades – affinity, dye–ligand and monoclonal antibody chromatography – cannot generally be used under conditions employing strong protein denaturants for obvious reasons. Hence one is left only with the older methods such as gel filtration (molecular sieving) which is broadly compatible with protein denaturants (though not all) and ion-exchange chromatography, which may be used in the presence of low ionic strength denaturants such as urea. The application of these two techniques can lead to a product of acceptable purity in some cases but further purification may well be required. Nevertheless, following purification, the product must be refolded into a native structure in order to regain the biological activity and although this may be fairly easy to achieve on a small scale it may present gross practical difficulties for scale up. Following protein refolding, or renaturation, and the maintenance of the protein as a soluble entity in water, it is possible to continue a purification scheme using the more powerful established methodologies discussed earlier. A final purification will be required in any case to separate successfully renatured molecules from those which failed to refold. Thus it would be useful to attempt the refolding and transfer of product from denaturant to aqueous solution at an early stage in a procedure.

An adequate explanation for the formation of inclusion bodies has not yet been forthcoming. From the current examples available in the literature it would appear that the problem of inclusion body formation is manifest during high level expression of eukaryotic genes in *E. coli.* There are few if any published exceptions to this pattern but it is worth emphasising that many of the types of proteins thus far cloned and expressed are hormones, e.g. insulin or receptor-binding proteins such as interferons and growth factors. Nevertheless, the normally soluble enzyme prochymosin also suffers from this problem (Marston *et al.* 1984). A further consideration is that the majority of these molecules are expressed as fusion proteins and one wonders whether this would influence the correct formation of product in the cells.

My own view of events is as follows: the high rate of protein synthesis leads to a high intracellular concentration of polypeptide. Rather than folding to the correct native structure, the protein chains coalesce with one another leading to the formation of an insoluble precipitate. Some intermolecular disulphide bridges could be formed within this precipitate. A proportion of the protein product will fold to the correct, active structure, although whether this is a significant quantity or not will depend upon the expression rate, the intracellular protein concentration, the nature of the protein itself and especially the rate at which newly synthesised chains fold to the tertiary structure. One can anticipate that the intracellular environmental conditions will also exert an influence on the folding of foreign proteins.

Why then does there not appear to be any difficulty in producing moderate to high levels of cloned *E. coli* proteins in *E. coli*? This question can only be answered rather speculatively at present, but these proteins are produced in their native environment and perhaps they have evolved to fold rapidly into their native structures following ribosomal synthesis. Considering the short generation times which occur during bacterial growth it follows that newly synthesised proteins must rapidly gain their native, functional structure to be of any value to the cell. Conversely, in eukaryotic cells there is unlikely to be the same degree of evolutionary pressure for optimal rates of protein folding. There is no fundamental difference between prokaryotic and eukaryotic proteins *per se*, although the latter may in many cases be glycosylated or subjected to higher degrees of post-translational modification. Importantly, one anticipates that cloned bacterial proteins will also form inclusion bodies if expressed intracellularly at a high rate and to a high level..

There may be solutions to these problems. Finer control over the rate of protein synthesis could pay dividends by reducing the intracellular concentration of folding polypeptide chains and thus producing the protein more gradually over a longer period of time. Expression of protein in other microorganisms such as yeast may well improve the yield of active product. The development of protein export systems using *Bacillus subtilis*, where the product is exported out of the cell, may alleviate some difficulties. The current trend of cloning into mammalian cells and producing product in tissue culture is giving encouraging, if expensive, results. Production by this means may be sufficiently economic for quantities of high value pharmaceutically active proteins, but for bulk production the problems dicussed above will need to be solved.

A recent and revolutionary advance has used protein engineering to overcome some of the downstream problems (Sassenfeld & Brewer 1984) associated with the production of urogastrone (the human analogue of mouse epidermal growth factor). Efficient product expression was obtained but the urogastrone was precipitated in the form of insoluble inclusion bodies within the cells – leading to all the difficulties outlined above.

To facilitate product recovery and purification, an additional gene sequence was added to the 3′ end of the gene which coded for an additional five arginine residues at the C-terminus of the protein. The new product, although hydrophilic, was still produced as an inclusion complex. However, by solubilisation in urea the urogastrone could be efficiently recovered on carboxymethyl ion-exchange resin and considerable purification effected by washing and high ionic strength elution (using urea solution as the solvent). Next the C-terminal fusion was removed using carboxypeptidase B, an exoproteinase specific for basic C-terminal residues, thus effecting a dramatic change in the iso-electric point of urogastrone. A second absorption/desorption CM ion-exchange purification yielded pure product since the contaminant bacterial proteins were essentially unchanged by carboxypeptidase treatment and still required high ionic strength for elution whereas the urogastrone was now eluted at low strength. Removal of urea and renaturation of the protein did not pose insurmountable difficulties in this particular case. This work is a good example of the types of benefit which will accrue through the use of protein engineering; in this case designing a protein for efficient downstream recovery and purification.

Notwithstanding the particular problems which still remain to be solved, the future for the application of recombinant DNA technology and protein engineering appears particularly attractive and will lead to new catalysts of great importance to industry.

REFERENCES

Blundell, T., Sibanda, B. L. & Pearl, L. (1983) *Nature* **304**, 273.

Harris, T. J. R. (1983) In: Williamson, R. (ed.), *Genetic Engineering*. Vol. 4, London Academic Press, p. 127.

Marston, F. A. O., Lowe, P. A., Doel, M. T., Schoemaker, J. M., White, S. & Angal, S. (1984) *Bio/Technology* **2**, 800.

Rosenberg, S., Barr, P. J., Najarian, R. C. & Hallewell, R. A. (1984) *Nature* **312**, 77.

Sassenfeld, H. M. & Brewer, S. J. (1984) *Bio/Technology* **2**, 76.

Wilkinson, A. J., Fersht, A. R., Blow, D. M., Carter, P., Winter, G., (1984) *Nature* **307**, 187.

12

Safety and regulatory aspects and the Inspectorate's approach

R. Nourish and N. J. Brace
Health and Safety Executive, F19, Occupational Hygiene Section (Microbiology), Stanley Precinct, Bootle, Merseyside L20 3QZ, UK

12.1 INTRODUCTION

Since the advent of genetic engineering some 13 years ago, there has been considerable interest in the practical application of the technique with its exciting possibilities for both improved production in the biotechnology industry and the creation of novel and economically important products. At the same time, there has been attendant concern about the potential hazards of this new technology which in turn has renewed interest in occupational hygiene aspects of the traditional fermentation industry. Particular attention has focussed on downstream processing, especially on those unit operations which utilise open or semi-contained plant since these have obvious potential for uncontrolled release of microbiological agents and products.

In addressing the subject of biotechnology in the United Kingdom, Government has stressed the importance of ensuring adequate and necessary measures for public and employee safety whilst at the same time recognising the need to avoid unduly restrictive controls on industrial development (HMSO 1981a).

This chapter is concerned with these occupational health and safety aspects. The existing legal requirements which apply to biotechnological processes and possible future developments are reviewed and the role of the Factory Inspectorate as an enforcement and advisory body is discussed. The general approach to control measures and plant containment is also outlined.

12.2 PERCEPTION OF RISK

The new biotechnology has progressed at a much slower speed than many predicted ten years ago and some people might point an accusing finger in the direction of safety requirements as one of the factors which may be restricting commercial exploitation and production in this industry. The finger might therefore also be pointed towards the Health and Safety Executive (HSE), although we hope that no one who has actually had dealings with HM Factory Inspectorate would feel that we were in some way over-zealously attempting to keep the gene genie in the bottle at all costs.

But, to put it quite bluntly, safety considerations could be an obstacle to progress unless they are properly addressed. There is a double-edged sword. On the one hand, as underlined by the joint ACARD/ABRC/Royal Society Report on biotechnology in 1980 (HMSO 1980) there are dangers of an over-reaction on safety and of excessive controls leading to possible constraints to development. On the other hand the price of under-reaction, perhaps one incident or an accident in a genetic manipulation laboratory (itself serious enough), could be a blot on the reputation of the conventional fermentation industry or even a serious obstacle to the development of the new biotechnology. This is particularly relevant in a society where, understandably, the emotive response to words such as genetic engineering, *E. coli* or lassa fever, legionnaires disease or AIDS is sometimes akin to that reserved for the nuclear industry. What is needed is a sense of proportion and a sense of balance. Indeed, in its white paper on biotechnology in 1981 the government charged those working in the field to maintain just such a balance – the balancing act of fostering innovation and commercial expedience whilst at the same time safeguarding employee and public safety.

As inspectors with special responsibilities for inspection of dangerous pathogen and genetic manipulation laboratories and industrial scale-up sites, we take this challenge very seriously indeed. The message is quite clear for industry and the Inspectorate alike. If the pitfalls associated with safety are to be avoided in the laboratory or production plant we have to aim for practical control measures that are consistent with the hazard.

At the heart of this business of getting the punishment to fit the crime is, of course, the vexed question of risk assessment. To date, it is true to say that the industry has taken this aspect very seriously and, following our contributions to previous conferences and meetings, we are pleased to say that inspectors have been invited to discuss in-house risk assessments and to comment on individual cases or projects at an early stage of their planning or development. It should be appreciated that inspectors do not contemplate these difficult issues of assessment and control in isolation. Our work is carried out in collaboration with colleagues in the microbiology policy section of the HSE Medical Division based in London. One of their tasks is to provide the secretariat to the two HSE Advisory Committees in this field. These committees – the Advisory Committee on Dangerous Pathogens (ACDP), and the newly formed Advisory Committee on Genetic Manipulation (ACGM) which replaced the Genetic Manipulation Advisory Group – are of course composed of employer and employee representatives as well as expert members. In essence, they advise Ministers, the

Health and Safety Commission and HSE generally on the hazards of work with pathogens and genetic manipulation and on standards for their control.

Of course, neither of these committees are responsible for on-site advice or inspection and these aspects together with law enforcement are the task of the Factory Inspectorate.

12.3 THE ROLE AND ORGANISATION OF THE FACTORY INSPECTORATE

During its 150-year history, the largest branch within HSE, HM Factory Inspectorate, has needed to adapt to the considerable changes in industrial technology, particularly in the last ten to fifteen years. In doing so, a pragmatic approach has been developed in establishing and applying the basic principles of the law and occupational hygiene to controlling health hazards. The Inspectorate's national and international reputation rests in part on its willingness to rationally debate and discuss risks and precuationary measures with all interested parties. The role of the Inspectorate is of course a blend of enforcement and advice. As specialist inspectors we attach particular importance to our advisory work in this field and to gain maximum benefit on both sides from this function we consider it important to at least outline our organisation and contact points in this chapter.

The Inspectorate employs area and specialist inspectors and in nearly all circumstances the former should be approached in the first instance when guidance is needed. These inspectors are based at 21 area offices throughout the country. When additional technical expertise is required they can call upon specialist engineering and microbiology inspectors. Within this area structure, a National Industry Group for chemical and pharmaceutical operations has been set up to ensure that there is a consistent approach towards standards throughout the country. This group is based in Merseyside and liaises with representatives of the industry.

The general exception to these working practices involves the specialist microbiology inspectors who are based at the Occupational Hygiene Section in Bootle. They should be approached directly for advice on:

(i) the most dangerous pathogen laboratories and industrial vaccine production units;
(ii) genetic manipulation laboratories and scale-up sites.

12.4 RELEVANT LEGAL REQUIREMENTS

The current regulatory framework which applies to this type of work consists of a package.

(i) The general principles which have been laid by statute, i.e. the Factories Act 1961 and the Health and Safety at Work etc. Act 1974.
(ii) Two sets of regulations which apply specifically to work with pathogens (HMSO 1981b) and genetic manipulation (HMSO 1978).
(iii) Codes of Practice which have been derived by the Advisory Committees.

The general legal framework governing operational aspects is provided by the Health and Safety at Work etc. Act 1974 (HSWA), which of course applies to all workplaces, and the Factories Act 1961. The key sections in the current context and indeed for all occupational hygiene considerations are section 63 of the Factories Act and section 2 of HSWA. In particular, section 2(2) of the latter act is part of the general duty imposed on the employer and requires:

> The provision and maintenance of the working environment for his employees that is, so far as is reasonably practicable, safe, without risks to health and adequate as regards facilities and arrangements for their welfare at work.

So in context, the environment of the downstream processing area must be without risk to health, but it should be noted that the duty is qualified by the words 'so far as is reasonably practicable'. This expression is not defined in the act but has acquired a clear meaning through established interpretation in the courts (Fife and Machin 1982). In a nutshell, the employer has to make a cost–risk analysis and assess on the one hand the risk of the work and on the other hand the difficulty, time, trouble and expense of the steps involved in avoiding the risk. If for example, the risks are very low and the costs and difficulties in controlling these are very high, then it might not be reasonably practicable to take those steps to avoid the risk. The greater the risk the less weight can be attached to the difficulties of avoidance. In the biotechnology industry, this risk may depend on the pathogenicity and virulence of the microorganism, toxicity of the microbial products and metabolites and the possibility of allergenic responses to protein and lipopolysaccharide material. On the basis of the risk assessment, some unit operations may not require specially contained plant whilst this may be necessary in other cases to ensure a healthy environment. For example, although there may be operational and maintenance difficulties, it is possible to design some downstream plant which prevents release of microbiological aerosols or to enclose the plant within a ventilated tent or enclosure. But it does not follow that these steps must always be taken irrespective of the risk.

The importance of local exhaust ventilation in controlling injurious or offensive dust, fume or other impurity at source is emphasised in section 63 of the Factories Act. This section encompasses the basic principles of occupational hygiene which are to reduce exposure to hazardous materials by containment methods and systems of work rather than reliance on personal protective equipment, such as respirators. Personal protection does have a part to play as a secondary measure and for use in emergencies, but the importance of proper selection, adequate maintenance and user training cannot be stressed too highly.

Currently there are no regulations which deal specifically with the operational aspects of the industrial or laboratory use of microorganisms. Both the Health and Safety (Genetic Manipulation) Regulations 1978 and the Health and Safety (Dangerous Pathogens) Regulations 1981 are only concerned with notification of the work.

The application of proposed new regulations to this field is currently under discussion (see p. 178) but the present statutory framework could be summarised as follows:

(i) The theme of the legal requirements is one of self-regulation and of getting the right balance. Moreover, the onus for risk assessment falls upon the employer at each and every stage of the laboratory work or production process. In performing its enforcement role, the Inspectorate checks that this assessment had been made conscientiously with consultation through the company's safety committee and representatives, and ensures that any necessary control measures are fully implemented.

(ii) The emphasis is on control of hazards at source rather than undue reliance on personal protection.

How are these legal principles translated into practical terms? First we need to distinguish the rather different approach to risk assessment and control which has been necessary in the area of biotechnology. The conventional chemical industry is well used to applying the concept of control limits or threshold limit values against which the results of environmental measurements can be judged. Compliance with control limits is an important indication, though not the only indication, on whether everything has been done 'so far as is reasonably practicable' to avoid worker exposure to toxic agents.

In many of the processes involved in biotechnology there are few relevant control limits and certainly there are none in the field of microbiology. Of course, that is not to say that microbes or microbial products are safe – merely that there is a lack of adequate information on which a quantitive approach can be based. In the absence of control limits what principles of assessment and control can be applied? This leads us to the work of the Advisory Committees and their Codes of Practice.

In the context of laboratory work, both the Genetic Manipulation Advisory Group (since replaced by ACGM) and the Advisory Committee on Dangerous Pathogens (ACDP) have specified containment levels directly corresponding to risk categories.

In the case of Genetic Manipulation Advisory Group (whose published guidance remains valid under the new arrangements) there are five categories (MRC 1981) whilst ACDP suggest four levels of containment in their report 'The categorisation of pathogens according to hazard and categories of containment' (HMSO 1984).

Although ACDP and GMAG have reached a similar end-point on the control of risks in their field, the method of risk assessment has been rather different. In genetic manipulation, the categorisation of recombinant DNA experiments are based on the assignment of risk values (negative exponents) for each of the factors: access, expression and damage. The product of these factors gives the overall risk value for the experiment which is then related directly to the appropriate containment level. Level 1 for example accords with a risk value of 10^{-12}. In contrast ACDP finally rejected such quantitative schemes because 'they could not be properly evaluated due to the lack of sufficiently reliable data'.

Instead the ACDP report, which relates to individual pathogens and not experiments, is based on an assessment of such data that was available by a concensus of experts. In producing the hazard classification and levels of containment it was recognised that there may in some cases be a need for local variation. Therefore a procedure is set out if it is proposed to use a level of containment lower than recommended by ACDP.

It is worth noting that the vast majority of the 185 manipulation centres in the country now work at the lowest level of containment (category 1) or under conditions of good microbiological practice. So, the current requirements of the regulatory framework hardly seem to present a problem in these laboratories.

How far can the principles of assessment and control in laboratories, be applied to work on an industrial scale? As regards the assessment process, it is first necessary to distinguish between the fermentation industry and the new biotechnology which employs organisms specifically engineered for industrial use. In the first case, although ACDP has not yet considered industrial operations specifically, it is suggested that the principles of assessment specified in the introduction to the code are applicable. For example, the local assessment of risk should take account both of the ACDP suggested hazard category and local factors such as virulence and transmissibility of the organism and the titre or concentration of the organisms in the broth to be handled. In addition, there is a need to consider allergenic and toxic properties of the organism or its products.

Clostridium tetani, for example, is categorised by ACDP as hazard group 2 on the basis of its infective hazard. But there is little doubt that a higher level of containment would be appropriate for large-scale growth taking into account the infectivity and more particularly the toxic hazards.

In the case of the new biotechnology, the potential risks deriving from scale-up work have been specifically considered.

GMAG note 12 (MRC 1982) points out that large-scale work differs from most laboratory work involving recombinant DNA in that the organism used and any product is completely defined prior to the start of production. In the factory the attention turns to potential genetic effects which are rare at the laboratory scale but which become significant when high concentrations or large volumes are handled. GMAG note 12 deals with additional factors to those of access, expression and damage that must be considered before proceeding to scale-up and production, e.g. frequency of integration of a plasmid into a host chromosome, survival of recombinants in the environment.

All scale-up proposals involving genetically manipulated organisms have to be notified to HSE under a voluntary notification scheme. Under these arrangements, procedures identified as low risk can go ahead after notification without awaiting comment and these comprise the majority of notifications received to date. Potentially higher risk procedures still require ACGM and HSE comment before work is begun. The practical arrangement is that ACGM comments primarily on the biological containment aspects whilst the Inspectorate deals with physical containment and site inspection. We accord a very high priority to these notifications so that there is a minimum of delay.

To date, the Inspectorate have dealt with scale-up proposals on a case by case basis. There are currently no published guidelines on containment aspects of industrial production and we would be concerned that any codes which may be produced in the future are sufficiently flexible to meet individual circumstances.

So far we have outlined the current regulatory framework in this country, but what of the future? The Health and Safety Commission's general approach to regulations and future strategy has been summarised (HMSO 1983): 'regulations necessary to deal with specific hazards and often also to implement European community directives will build upon the comprehensive obligations of the HSW Act and will require what is necessary, but no more than is necessary to set the right regulatory framework to control specific hazards'.

The idea is that most of the detail can be dealt with in approved codes of practice, which can be more readily updated and amended, or in guidance developed in close consultation with those directly affected, for instance through industrial advisory committees. The message is clear – there is no intention to move towards very detailed regulations which soon become outpaced by technology.

Recently the Health and Safety Commission has published a consultative document on proposed regulations on the control of substances hazardous to health (HSE 1984). In the consultative draft regulations the proposal is made that pathogens be encompassed by the regulations. The ACDP has been asked whether the draft regulations are an appropriate framework for this and if so, to advise on the format of a specific pathogens approved Code of Practice. Clearly, the views of ACGM and industry will also be sought and taken into account.

Even though we have emphasised that any form of regulation or code of practice would be general in nature, from our own experience it certainly appears that designers and production staff would welcome some form of technical guidance on practical control measures. We hope to meet this need following project work in the workplace and discussion with people in the field.

12.5 THE INTERNATIONAL PICTURE

A group is currently conducting a study on safety and regulations in biotechnology under the auspices of the OECD committee for scientific and technological policy. The study is concentrating on a review of country positions on the safety in use of genetically engineered organisms and the identification of scientific criteria which could underlie guidelines for safe handling. It is to explore possible ways and means of monitoring future production and use and it is hoped that this will be a step towards better international harmonisation of guidelines, codes of practice or regulations.

12.6 CONTROL MEASURES

It is fair to say that potential health problems caused by the release of microorganisms and products from downstream processing plant is well acknowledged by the industry itself. Cases have been reported and also details of control measures which have been specifically adopted have been described (Mayes 1982; Ekenvall *et al.* 1983; Newhouse *et al.* 1970; Juniper *et al.* 1977; Harris-Smith & Evans 1974; Dunnill 1982; Lawrence & Barry 1982; Hemert 1982; East *et al.* 1984).

In some countries, standards have already been introduced for scale-up and production plant utilising genetically manipulated organisms (NIH 1983; DST 1984). They have been based on clearly defined containment levels which are directly linked to laboratory standards. The approach in this country has been to match the control measures to the hazards involved at each unit operation following a case by case risk assessment. Emphasis has been placed on control at source by good plant integrity and local exhaust ventilation. This is certainly not a novel idea (Rose & Nourish 1982; Nourish 1983) but embodies traditional occupational hygiene principles which have been applied for many years to toxic substances and dusts.

As always, elimination or substitution of the hazard are primary objectives, particularly when considering the capital, maintenance and operational costs of running a plant with a standard of containment required for high risk organisms and products. If possible, it is advisable to select a safer substitute organism or strain. Likewise, elimination of a particularly hazardous step may enable a lower standard of containment to be adopted at subsequent stages, e.g. killing the organisms, dextoxification of toxin products.

With the exception of small-scale, high-speciality products, in our view only two levels of plant containment are likely to be used by the production industry. The first could be described as primary containment with plant which is designed to prevent release, e.g. double mechanical seals on impeller shafts, steam sterilisable ports. With the exception of features which might be additionally required in certain cases, e.g. HEPA standard filtration on exhaust gases, these are no more than necessary to achieve aseptic operation. On a related theme, the British Occupational Hygiene Society has recently published guidance of fugitive emissions which contains information on values, seals etc. (Science Reviews 1984).

A higher level of containment, i.e. secondary, would be appropriate if release is likely owing to the design or failure in plant integrity. This standard of containment therefore involves the use of local exhaust ventilation, glove boxes or even safety cabinets and flexible film isolators for small items of plant. A less desirable option is achieved by locating the plant in a ventilated room where the operation is remotely controlled and personnel only enter whilst wearing respiratory protective equipment.

Having selected and provided the appropriate containment devices, it is important that these are performance tested, e.g. checks for plant leakage and filter penetration.

Many of the general safety precautions adopted for laboratory-scale work are also necessary at the industrial level, i.e. operator training, good technique, supervision and medical monitoring.

12.7 CONCLUSIONS

The agency which exercises enforcement in this country, the Health and Safety Executive, is also very closely involved in standard setting and derivation of codes of practice. We reckon that this is likely to produce standards which are both pragmatic and enforceable.

It is fair to say that good professional relationships have built up with trust on both sides between the Inspectorate and industry. We have established and maintained close contacts with the developing biotechnology industry and this has led, for example, to collaborative project work on risk assessment and control in the workplace.

In our opinion, in most cases it will be more practicable to assess and take appropriate control measures at each unit operation rather than the complete process. This approach is certainly not original.

Within the Health and Safety Executive, we very much welcome the opportunity to get involved early in proposals for scale-up and production as well as discussion about the technologies involved. Healthy discussion and debate can only help to ensure that the right balance is obtained. We want to be involved in sharing, understanding and preventing any problems.

Finally, we understand what is meant by the term confidentiality.

REFERENCES

DST (1984) Recombinant DNA Monitoring Committee, Publication No. 4, Guidelines for large scale work with recombinant DNA, Department of Science and Technology, Australia.

Dunnill, P. (1982) Biosafety in the large-scale isolation of intracellular microbial enzymes. *Chem. Ind.* **22**, 877–879.

East, D., Stinnett, T. & Thoma, R. W. (1984) Reduction of biological risk in fermentation processes by physical containment. *Dev. Ind. Micro.* **25**, 89–105.

Ekenvall, L., Dolling, B., Gothe, C.-J., Ebbinghaus, L., Von Stedingk, L.-V. & Wasserman, J. (1983) Single-cell protein as an occupational hazard. *J. Ind. Med.* **40**, 212–215.

Fife, I. & Machin, E. A. (1982) *Redgrave's Health and Safety in Factories.* London: Butterworths Shaw and Sons.

Harris-Smith, C. P. & Evans, C. G. T. (1974) Bioengineering and protection during hazardous microbiological processes, Symp. No. 4, 837–855.

Hemert, P. van (1982) Biosafety aspects of a closed system Westfalia-type continuous centrifuge *Chem. Ind.* **22**, 889–891.

HMSO (1978) Health and Safety (Genetic Manipulation) Regulations, SI 1978 No. 752.

HMSO (1980) Advisory Council for Applied Research and Development, Advisory Board for Research Councils, Royal Society, Joint Report, Biotechnology.

HMSO (1981a) Government White Paper, Biotechnology, CMND 8177.

HMSO (1981b) Health and Safety (Dangerous Pathogens) Regulations, SI 1981 No. 1011.

HMSO (1983) Health and Safety Commission, Plan of Work 1983–84 and onwards.

HMSO (1984) Advisory Committee on Dangerous Pathogens, Categorisation of pathogens according to hazard and categories of containment.

HSE (1984) Health and Safety Commission. Control of Substances Hazardous to Health – draft regulations and draft approved codes of practice.

Juniper, C. P., How, M. J., Goodwin, B. F. J. & Kinshott, A. K. (1977) *Bacillus subtilis* enzymes: a seven year clinical, epidemiological and immunological study of an industrial allergen. *J. Soc. Occup. Med.* **27**, 3–12.

Lawrence, A. & Barry, A. (1982) Potential hazards associated with the large-scale manufacture of bacterial vaccines. *Chem. Ind.* **22**, 880–884.

Mayes, R. W. (1982) Lack of allergic reactions in workers exposed to Pruteen (bacterial single-cell protein). *B. Journ. Ind. Med.* **39**, 183–186.

MRC (1981) Genetic Manipulation Advisory Group. GMAG notes 14–15, Medical Research Council, London.

MRC (1982) Gentic Manipulation Advisory Group. GMAG Note 12 (revised), Medical Research Council, London.

Newhouse, M. L., Tagg, B., Pocock, S. J. & McEwan, A. C. (1970) An epidemiological study of workers producing enzyme washing powders. *The Lancet* 4 April, 689–693.

NIH (1983) National Institutes of Health, Guidelines for Research involving recombinant DNA molecules. USA Federal Register 48, 106, 24556–24581.

Nourish, R. (1983) Safety in Biotechnology: the role of the enforcement agency in the UK, the HSE approach, Biotech 83 – proceedings of the international conference on the commercial applications and implications of biotechnology. Northwood, UK: On-line Publications, pp. 177–188.

Rose, T. J. & Nourish, R. (1982) Occupational hygiene considerations – an HSE viewpoint. *Chem. Ind.* **22**, 892–894.

Science Reviews (1984) British Occupational Hygiene Society, Technical Guide No. 3, Fugitive emissions of vapours from process equipment.

13

Whole broth solvent extraction

K. H. Brunner
Westfalia Separator AG, D-4740 Oelde 1, FRG

13.1 INTRODUCTION

For many applications extraction has proved a gentle, economical process for separation of two mutually dissolved substances. In the pharmaceutical industry the special requirements for recovery of antibiotics from fermentation broths led to the development of centrifugal extractors at the end of the 1940s. It is only because of the rapid mass transfer during the comparatively short contact time of the phases involved in the centrifugal field that yields are achieved which make economic production possible at all.

13.2 PROBLEMS OF THE PROCESS

Large-scale production of antibiotics has hardly changed during the last three decades and consists essentially of the following processing stages (Fig. 13.1): fermentation; extraction; crystallisation (precipitation).

During a fermentation process lasting about a week, the antibiotic is formed by specific microorganisms dispersed in a nutrient medium. It is then present in a dissolved state or partially combined with the solids and must be isolated from the nutrient medium as quickly as possible. For this procedure all solid mycelium and nutrient medium components must be separated out, as otherwise, clogging and yield losses would occur in standard design centrifugal extractors, which are designed for purely liquid–liquid extraction.

After suitable clarification of the fermentation broth in rotary vacuum filters and centrifuges, in many cases a readjustment of the pH value follows,

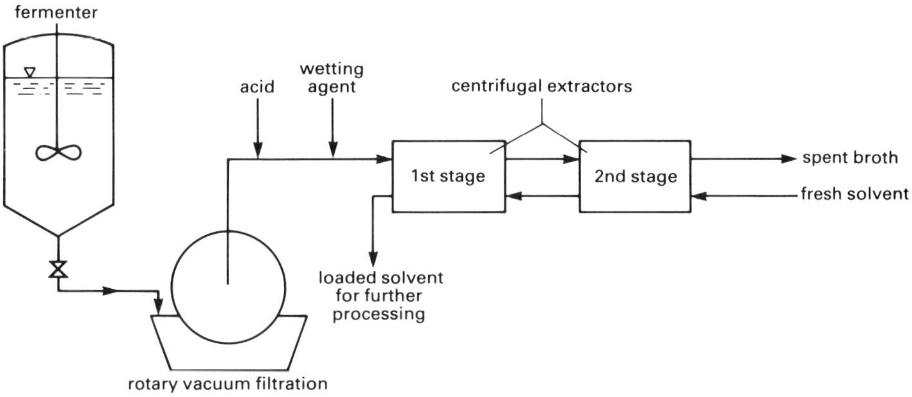

Fig. 13.1 – Flow sheet of the recovery of antibiotics (i.e. penicillin).

immediately before extraction, which causes for example in the case of penicillin, flocculation of protein.

The finely dispersed protein particles form deposits in the extractors, cause losses in yield and give rise to impurities in the raffinate and extract. Certain cleaning intervals are necessary, so that the solids separated out can be removed.

At the end of this process, further treatment of the enriched solvent by means of re-extraction and crystallisation is carried out. However, this process will not be discussed here. Process inadequacies and problems of extraction were the reasons for the following developments.

Particularly in the case of extraction of secondary metabolites combined with the mycelium, it was necessary to bring the solids into contact with the solvent to avoid losses as far as possible. Corresponding work was carried out to find the most suitable apparatus for this purpose. (Schlerfler 1979, Katinger *et al.* 1981.)

For solids contents of up to 60% by volume of separable mycelium and nutrient components, the use of centrifugal decanters with intermediate centrifugal mixers seemed to be the most suitable method. In a two-stage countercurrent system, adequate solvent extraction yields of over 95% were attainable.

13.3 EXTRACTING DECANTERS

Parallel to this work, the extracting decanter (German patent No. 2,701,763) was developed. Figure 13.2 shows a schematic drawing of this machine. It differs from the standard decanter mainly in the separate feeds of the two phases and in the construction of the transporting scroll. The two phases are fed into the rotating bowl at different points. Contact of the phases takes place in countercurrent mode in the inlet zones and over a distance of 3 metres in the spirals of the scroll.

The separating disc at the transition from cylindrical to conical bowl section blocks the light phase which flows into the centripetal pump chamber over a ring dam at the cylindrical end of the bowl, from where it is drawn off by the stationary centripetal pump.

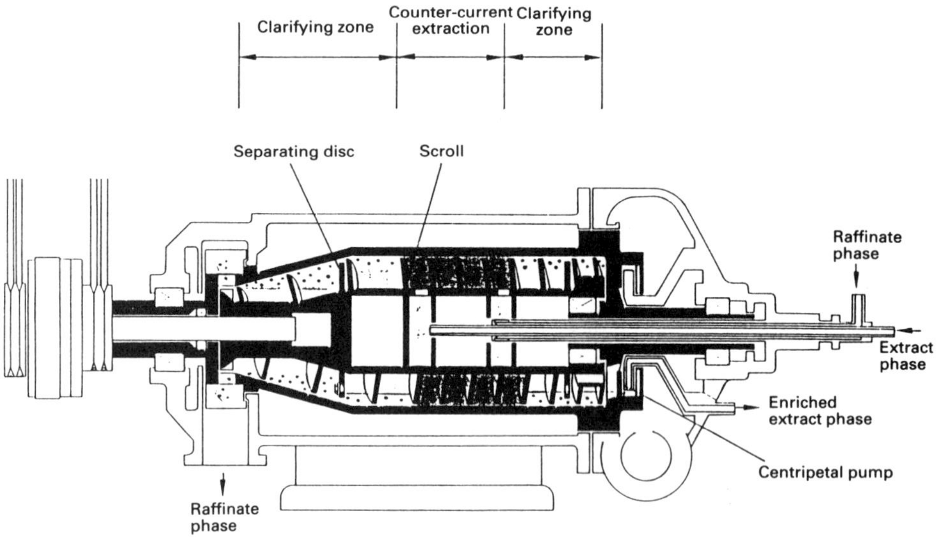

Fig. 13.2 – Decanter for continuous counter-current extraction.

The heavy (aqueous) phase, enriched with solids, discharges freely from the conical bowl end. Sedimented solids are collected by the scroll and continuously discharge from the rotating bowl. The separating zone between the two phases depends on the density difference and on the overflow diameters at the conical and cylindrical ends of the bowl and, for optimum phase separation, is located approximately in the middle of the total liquid pool.

13.4 PROCESS IMPROVEMENTS

Use of this extracting decanter has resulted in a series of improvements in the previously described process of antibiotics recovery. The appropriate processing sequence is illustrated in Fig. 13.3.

As the fermentation broth can be fed into an extraction process without previous filtration, there are consequences for continuity of the process and for considerable savings in capital costs of building conversions, space, apparatus and tanks. Frequent complaints of difficulties with the filterability of so-called 'bad' or partly infected fermenters are a thing of the past.

Furthermore, the whole broth extraction gives an increase in the yield of active substance of up to 10%, which is otherwise combined with the filter cake and is lost if the washing is inadequate. The wash water necessary in the conventional process increases the load on the extraction process with diluted

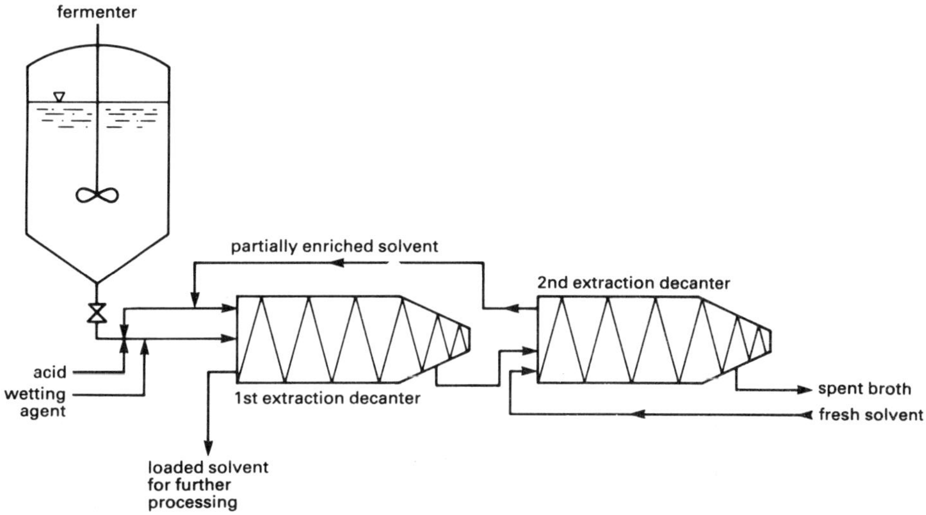

Fig. 13.3 – Flow sheet of direct extraction process (i.e. penicillin).

culture filtrate and solvent extract and increases the volume of waste aqueous phase at the end.

In the processing cycle introduced here, the full contents of the fermenter are fed directly to the first extracting decanter where they are brought into contact with the already slightly enriched solvent from the second decanter stage. In some cases, it is necessary to add a wetting agent or to mix part of the solvent with the broth before the pH value is adjusted.

The fermentation broth from the first extracting decanter is then fed to the second stage and leaves this decanter after further extraction with fresh solvent. The final steps are azeotropic distillation, to remove the solvent combined with the broth, and further biological and mechanical treatment of the waste aqueous stream.

The solvent, enriched with active substance, drawn off from the first stage, can be subsequently clarified and washed in a centrifugal clarifier if required. Then, depending on quality requirements, further purification or direct crystallisation from the solvent is carried out.

13.5 SEMI-INDUSTRIAL EXTRACTION TESTS

Industrial operation testing of direct extraction was carried out on a semi-industrial scale in several plants producing antibiotics. The pilot plant used for this purpose is shown in Fig. 13.4. It consists essentially of two counter-current extracting decanters (CA 226–290), which are mounted on a common foundation frame with all necessary pumps, metering devices, tanks and pipelines. The technical data of the decanters are summarised in Table 13.1.

Fig. 13.4 – Extraction pilot plant (2 × CA 226–290).

Table 13.1.

Bowl speed	5100 rpm (3200 g)
Differential scroll speed	35 rpm
Bowl diameter	220 mm
Bowl length	750 mm
Half cone opening angle	15°
Throughput capacity (fermentation broth)	1–1.5 m^3 h^{-1}
Total throughput capacity (broth + solvent)	1.3–2 m^3 h^{-1}
Connected load	11 kW

With this pilot installation, penicillin G, penicillin V, erythromycin and pleuromulin were extracted. The solvent used in most cases was butyl acetate; in one case methyl isobutyl ketone was tested. Separable mycelium and nutrient medium in the fermentation broth were in the range 10–60% by volume.

13.5.1 Direct extraction of penicillin

For extraction of penicillin G and V, it was necessary to feed in part of the solvent before acidification in the first stage, so that the immediately produced penicillanic acid could be absorbed by the organic phase. Consequently, pure counter-current extraction was no longer possible in the first decanter stage. However, no unfavourable effect on the extraction yield could be detected.

Phase separation in the extracting decanters could be decisively influenced by addition of wetting agent. With an addition of 0.2–0.5% dissolved Demulso or Nadar Cento surfactant in the butyl acetate, throughputs of up to 1800 litres per hour were attainable. The discharge of solvent from the first decanter stage was then almost free of impurities. However, a subsequent clarification stage with a centrifugal clarifier can be provided as a precautionary measure.

The spent fermentation broth discharged by the second stage decanter contained about 3–5% butyl acetate in the solids, which can be recovered by heat treatment after extraction. With a phase ratio of 2:1 to 3:1 (broth/solvent), extraction yields from the fermentation broth of 97–98% were attained. The total throughput capacity varied between 1.2 and 2.0 $m^3 h^{-1}$, without a detectable effect on the extraction yield or separating efficiency.

The quality of the end-product thus recovered was then tested by various methods and compared with the pencillin produced in the main plant. No significant differences in colour value, specific units or other quality-influencing parameters could be found. It can thus be guaranteed that direct extraction from the culture broth does not produce any change in selectivity, and the quality of the end-product remains unchanged.

Comparison of the yields from direct extraction with those of the existing industrial installation, consisting of vacuum rotary filters and Podbielniak extractors, showed a clear average increased yield of pure penicillin of at least

5%. Consequently, in addition to the advantage of improved process continuity, the increased yield means considerably higher profit margins.

13.5.2 Direct extraction of erythromycin

Further proof of the superiority of the direct extraction process was provided by pilot tests with erythromycin. The only difference here was the mass transfer from the basic medium and the use of Aquad surfactant for improvement of phase separation.

In a plant with vacuum filtration and an ABE type 216 extractor, evaluation of 11 comparison tests showed average increases in yield of 10–15% pure erythromycin. In order to arrive at the most accurate result possible, in each test 5 m^3 of fermentation broth was processed and, at the end, the total amounts of pure erythromycin were compared.

Because of the structure of mycelium and nutrient solution components, up to 15% of the erythromycin is lost when filtration is employed, despite washing of the filter cake. The compound is extracted with no great difficulty with direct phase contact. The throughput capacity of the pilot installation was in most cases, here also, around 1.5 $m^3 h^{-1}$ of broth, and the phase ratio (broth: butyl acetate) was 3:1 to 4:1. Phase separation in the decanters was so good that it was possible to dispense with post-clarification of the solvent. Because of the relatively high solids content of the fermentation broth, there is absorption of 3–4% solvent which cannot be mechanically separated out and must be recovered by distillation after the extraction process.

To sum up, the tests with erythromycin established that the mycelium components, which are far more difficult to separate out than those in penicillin broths, present no problems in direct extraction, and satisfactory yields can also be obtained with the two-stage extracting decanter plant.

13.5.3 Influence of wetting agents

The influence of nine different wetting agents on the yield and separation efficiency were studied, and the corresponding results are shown in Fig. 13.5. An optimum wetting agent should give the best phase separation with highest yield at lowest consumption possible. The results of a long-term trial on penicillin V with the wetting agent called Genamin TAB 100 are given in Fig. 13.6.

A comparison of a counter-current extraction with two stages and three stages shows a slight increase of the yield by 1–2% which already pays for this extra stage. An average yield of between 97% and 99% could be obtained, and the consumption of wetting agent was approximately 3 kg m^{-3} broth. Although different strains were processed during the trial period, no significant variations of extraction yield or solvent losses were detected during the test period.

13.5.4 Consideration of economics

The financial advantages of the new technology can be demonstrated by a consideration of economics. In penicillin plants of average size the contents

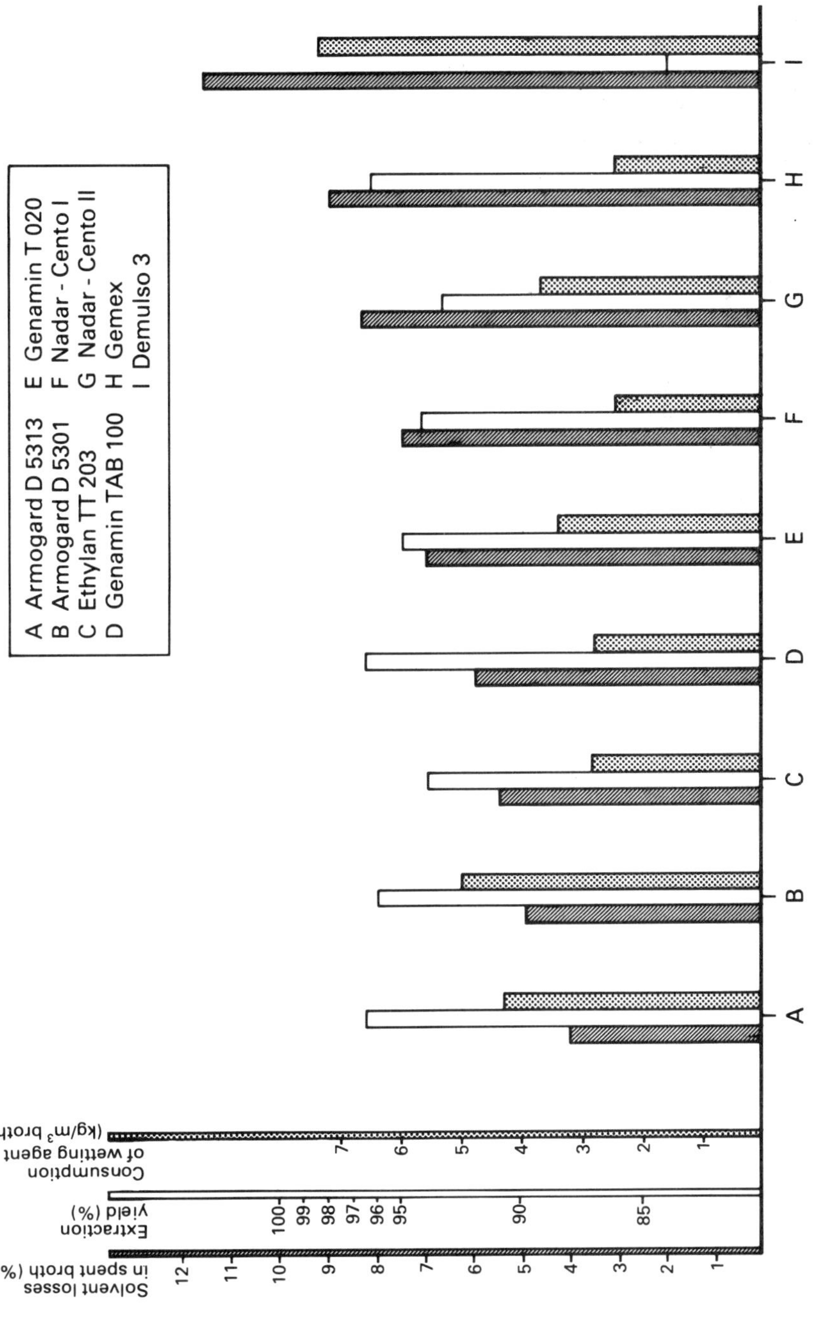

Fig. 13.5 – Test results with different wetting agents.

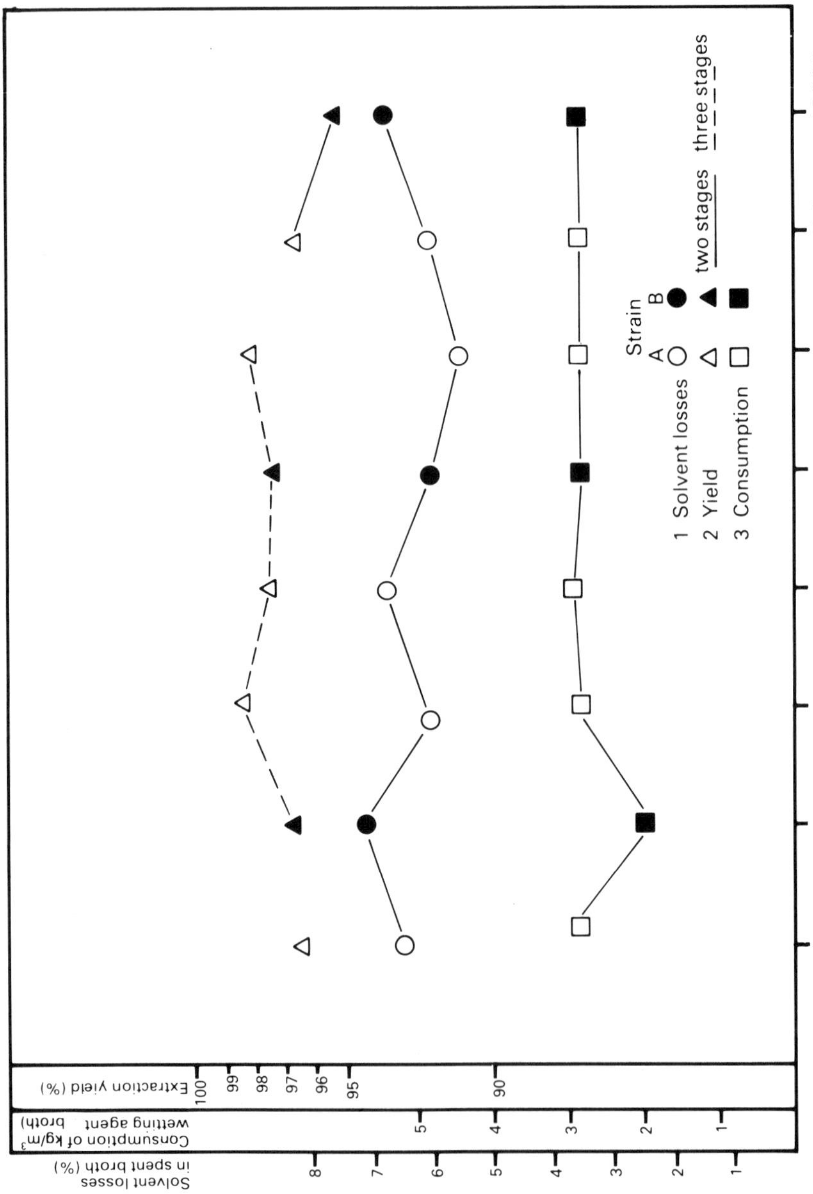

Fig. 13.6 – Long-term industrial test results with penicillin V.

of one approximately 120 m^3 fermenter are processed per day. A concentration of about 25 g dm^{-3} results in a yield of 3.1 tons of raw penicillin from one fermenter. The corresponding annual capacity is thus about 1000 tons penicillin. With present-day recovery methods a loss in yield of 5–15% by instability, filtration and extraction cannot be avoided.

In the case of direct extraction, an extra yield of at least 5% or 50 tons raw penicillin per year can be obtained. When calculating with a price for raw penicillin of about 20 US\$ kg^{-1}, US\$ 1 million can be gained only by extracting the whole broth with the counter-current extraction decanter. This does not even take into account the savings of filter aid.

As the investment for the main equipment of a plant for the extraction of 12 m^3 h^{-1} of fermentation broth amounts to about half a million US\$, the plant can be amortised within one year. In addition to technological superiority there are also considerable economic advantages to the new process.

13.6 LABORATORY TESTS WITH A TOLUENE/WATER/ACETONE SYSTEM

In order to establish with reasonable certainty the number of theoretical stages achievable in an extracting decanter, laboratory tests were carried out with a toluene/water/acetone system. Based on the results, an accurate scale-up and also a comparison with centrifugal extractors of other design was possible.

A counter-current extracting decanter, with the technical specification detailed in Table 13.1, from the pilot installation already tested in several antibiotics plants, was used. This machine was employed to test the extent to which the extraction yield could be affected by varying the differential speed of the scroll, the throughput capacity or the phase ratio. The initial concentration of acetone in toluene was in all cases about 10% by weight. Mass transfer was from the toluene into water and the processing temperature was a constant 18°C.

With a fixed bowl speed of 5100 rpm and a phase ratio of water to toluene of 2:1 for all tests, first the time for attainment of a constant extraction yield was examined. From the curve in Fig. 13.7 it can be seen that after a period of operation as short as one minute, equlibrium corresponding to 1.76 theoretical stages was attained. With other centrifugal extractors this time can be up to 20 minutes. Surprisingly, the yields measured at the beginning are here greater than those of the stationary condition, in contrast to other centrifugal extractors.

In this preliminary test the time to attain equilibrium sampling was set at 2 minutes for the ensuing series of measurements. Subsequent variation of throughput in the range 0.9–1.8 m^3 h^{-1} at a phase ratio of 2:1 (water/toluene) always led to a uniformly good yield, corresponding to 1.7 theoretical stages. The feed concentration was 9.68% by weight acetone in toluene. In the discharging toluene, 1.4% by weight acetone was still detectable by means of titration.

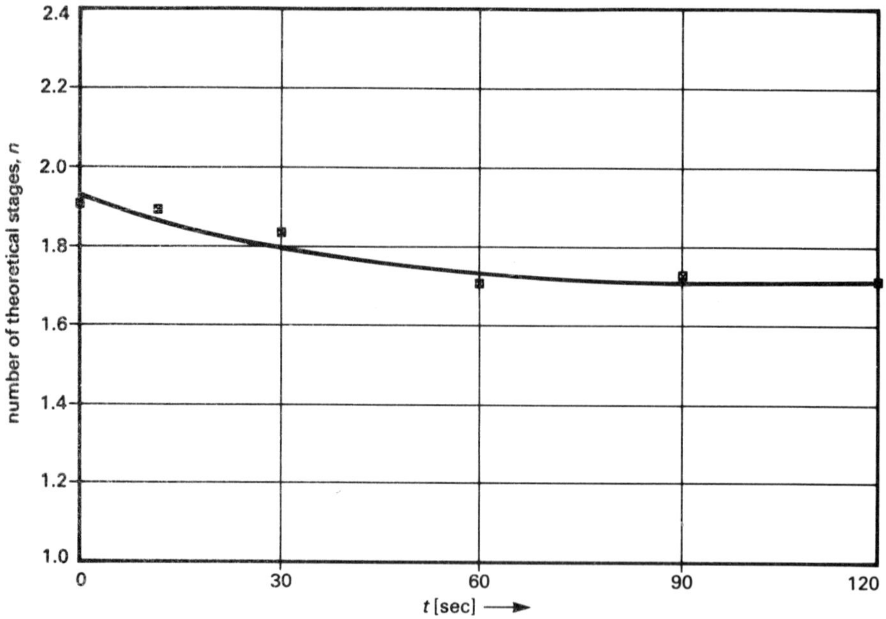

Fig. 13.7 – Time pattern until stationary extraction is reached.

It can also be deduced from Fig. 13.8 that at various phase ratios there is no effect on the efficiency of the extracting decanter. The total throughput was, indeed, varied slightly in the range 1.25–2.0 $m^3 h^{-1}$ at the same time; however, according to previous measurements the effect of throughput can be precluded.

As a final variable, the differential speed of the scroll was examined. The transporting scroll is obviously not important for transporting solids in a toluene/water/acetone system; however, the differential speed of the scroll, due to its stirring effect, influences the separation and extraction efficiency.

As was evident from corresponding tests, the extraction yield indeed rises with increasing differential speed; however, separation efficiency is reduced at a differential speed of 85 rpm and a total throughput capacity of 1.5 $m^3 h^{-1}$. The optimum differential scroll speed of 45 rpm was found in this way. Installation of further agitator elements at this point to provide intensive mixing in the contact zone is a possibility. However, with this modification it was repeatedly found that separating efficiency is impaired and extraction yield is only moderately increased.

The results obtained with a toluene/water/acetone system in an extracting decanter can be summarised as follows. The extraction yield attainable in a machine is virtually independent of all variables, such as throughput, differential speed of scroll and phase ratio, and corresponds to a yield of about 1.7 theoretical stages.

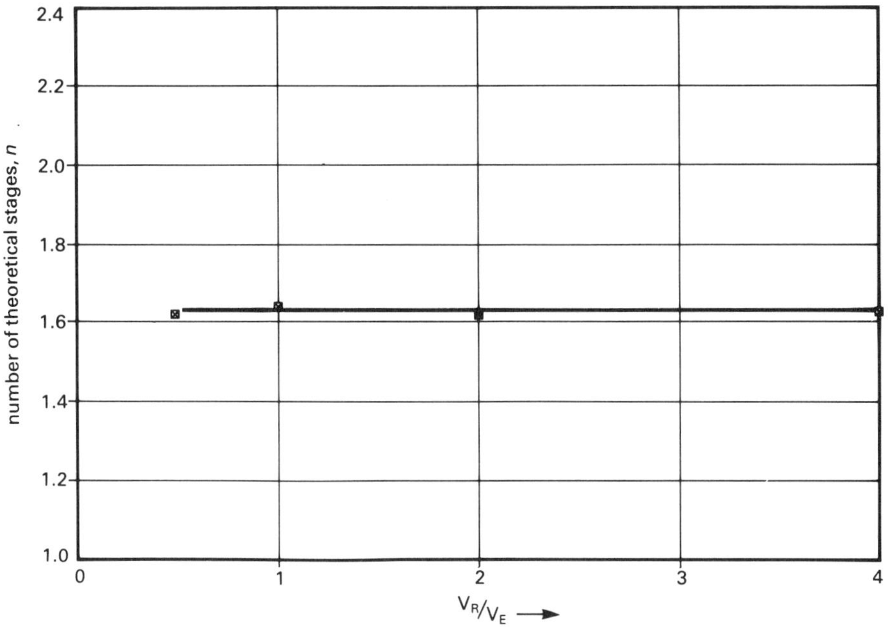

Fig. 13.8 – Number of theoretical stages in dependence of the phase ratio.

13.7 SUMMARY

These results are confirmed by the trials with penicillin and erythromycin on a semi-industrial scale, in which yields were obtained, in two counter-current extracting decanters, which suggest an efficiency of three to four theoretical stages. The decisive advantage of this new extractor, in addition to the constancy of the number of theoretical stages, is its ability to cope with solids. Particularly in the case of extraction of antibiotics from fermentation broths, solids collect which are often difficult to separate out and frequently combine with valuable substances. Centrifugal extractors, familiar up to now, are not suitable for whole broth extraction, so that filtration of the broth is essential. A great deal of space is required for the rotary vacuum filters necessary for the process, and sanitary and filtration problems are unavoidable. Despite washing of the filter cake, yield losses of 5–15% are common.

All these disadvantages of the methods of antibiotics recovery from fermentation broths in use for about the last 30 years led to the development of the counter-current extracting decanter described here. As has been shown by the semi-industrial extraction trials, a more economical method of recovering antibiotics has been successfully developed.

MANUFACTURERS OF WETTING AGENTS

Akzo Chemie	: Armogard D 5313
Littleborough, U.K.	: Armogard D 5301
	Arquad
British Petroleum, Hythe, U.K.	: Gemex
Diamond Shamrock, Leeds, U. K.	: Ethylan TT 203
Höchst AG, W. Germany	: Genamin TAB 100
	Genamin T 020
Nadar Chimica, Milan, Italy	: Nadar Cento
Petrolite Ltd., Liverpool, U.K.	: Demulso

REFERENCES

Brunner, K. H., Scherfler, H., & Stölting, M., (1981) Experiences with the countercurrent extraction decanter in the recovery of antibiotics from fermentation broths. *Verfahrenstechnik* No. 15, 619–622.†

Brunner, K. H., Scherfler, H., & Stölting, M., (1981) Centrifugal decanter of special design for direct extraction processes. Lecture at the 7th International CHISA Congress, Prague.†

Katinger, H., Wibbelt, F., Scherfler, H., (1981) Continuous direct extraction of antibiotics by means of centrifugal separators. *Verfahrenstechnik* No. 3, 179–182.

Scherfler, H., (1979). Continuous direct extraction of antibiotics. Thesis, Vienna.

† These references do not appear in the text.

Index